SpringerBriefs in Molecular Science

For further volumes:
http://www.springer.com/series/8898

Saima Parveen · Muhammad Sohail Aslam
Lianzhe Hu · Guobao Xu

Electrogenerated Chemiluminescence

Protocols and Applications

 Springer

Saima Parveen
Lianzhe Hu
Guobao Xu
State Key Laboratory of Electroanalytical
  Chemistry
Changchun Institute of Applied Chemistry
Chinese Academy of Sciences
Changchun
People's Republic of China

Saima Parveen
Lianzhe Hu
Guobao Xu
University of the Chinese Academy
  of Sciences
Beijing
People's Republic of China

Saima Parveen
Department of Chemistry
Faculty of Science
The Islamia University of Bahawalpur
Bahawalpur
Pakistan

Muhammad Sohail Aslam
University College of Pharmacy
University of the Punjab
Lahore
Pakistan

ISSN 2191-5407   ISSN 2191-5415   (electronic)
ISBN 978-3-642-39554-3   ISBN 978-3-642-39555-0   (eBook)
DOI 10.1007/978-3-642-39555-0
Springer Heidelberg New York Dordrecht London

Library of Congress Control Number: 2013943564

Printed on acid-free paper

Springer is part of Springer Science+Business Media (www.springer.com)

This work is dedicated to my
Parents, Brother, Sisters, and my Husband

Saima Parveen

Preface

Electrochemiluminescence techniques with remarkably high sensitivity and extremely wide dynamic range have hundreds of millions of dollars in sales per year. This technique is extensively used in clinical analysis and scientific research. The book *Electrogenerated Chemiluminescence: Protocols and Applications* is designed as a brief and concise introduction to electrochemiluminescence. The book is committed to furnish the idea of processes, theory, and the recent developments involved in the instrumentation. Another chapter devoted on the coupling of electrogenerated chemiluminescence covers essentially many different techniques combined with electrochemiluminescence in order to attain greater sensitivity and selectivity. Miniaturization/micro total analysis system (μTAS) is an emergent field and is gaining a prospective importance. It is also crucial and important to know something of the language of electrochemiluminescence in terms of miniaturization/μTAS. So we have added a nice approach to view the concepts of the topic. The chapter *Applications of Electrochemiluminescence* is focused on the great potentiality of ECL in various fields.

Our goal in preparing this book is to provide more appropriate and enough understanding of the fundamental concepts about the world of electrochemiluminescence to allow reader to gain knowledge while turning the pages.

This project was kindly supported by the National Natural Science Foundation of China (No. 21175126), Chinese Academy of Sciences (CAS), and the Academy of Sciences for the Developing World (TWAS). The authors would also like to acknowledge Higher Education Commission, Pakistan.

Saima Parveen
Muhammad Sohail Aslam
Lianzhe Hu
Guobao Xu

Contents

Abbreviations

TPA	Tripropyl amine
TBA	Tributyl amine
TEA	Triethyl amine
PMT	Photomultiplier tube
LEDIF	Sequential light-emitting diode-induced fluorescence
AAs	Amino acids
NDA	Naphthalene-2,3-dicarboxaldehyde
pHPP	p-hydroxyphenylpyruvic acid
Eu-PB	Europium(III)-doped prussian blue analogue
MH	Metformin hydrochloride
CPE	Carbon paste electrode
PDMS	Poly(dimethylsiloxane)
RSD	Relative standard deviation
GOD	Glucose oxidase
PTFE	Polytetrafluoroethylene
GDH	Glucose dehydrogenase
Chx	Choline oxidase
DEAE	(Diethylaminoethyl) Sepharose
ABEI	N-(aminobutyl)-N-ethylisoluminol
hIgG	Human immunogloblin G
ROSs	Reactive oxygen species
LPA	L-phenylalanine
MISPE	Molecularly imprinted solid-phase extraction
Chit	Chitosan
DA	Dopamine
CSF	Cerebro-spinal fluid
SPCE	Screen-printed carbon electrodes
SIA	Sequential injection analysis
TNT	2,4,6-trinitrotoluene
PE	Perylene
PSS	2,4-dichlorophenoxyacetic acid (2,4-D) Poly(p-styrenesulfonate)
CPZ	Chlorpromazine
HA	Humic acid

ChOx	Choline oxidase
ACPG	Aminopropyl-controlled pore glass beads
TCs	Tetracyclines
OTC	Oxytetracycline
COD	Cholesterol oxidase
$(Ru(bpy)_3{}^{2+})$	Tris(2,2′-bipyridine)ruthenium (II)
DPH	Diphenhydramine
EPH	Ephedrine
CV	Cyclic voltammetry
(SERS)	Surface-enhanced Raman scattering
PL	Photoluminescence
PPZ	Perphenazine
CCE	Cement carbon electrode
MG	Malachite green
DDQ	2,3-dichloro-5,6-dicyano-1,4-benzoquinone
LMG	Leucomalachite green
SPE	Solid phase extraction
AMPA	Aminomethylphosphonic acid
(MSPD)	Matrix solid phase dispersion
poly-β-CD	Poly-β-cyclodextrin
LODs	Limits of detection
HME	Heating cylindrical microelectrode
DPZ	Dioxopromethazine
CZE	Capillary zone electrophoresis
DCPE	Dual-cloud point extraction
IAA	Indole-3-acetic acid
IBA	Indole-3-butyric acid
SEB	Staphylococcal enterotoxin B
SDS	Sodium dodecyl sulfate
HITOE	Heated ITO electrode
DMAET	(Dimethylamino)ethanethiol
BSA	Bovine serum albumin
Fc	Ferrocene
Fc-MB	Ferrocene-labeled molecular beacon
FcA	Ferrocenemonocarboxylic acid
EPR	Electron paramagnetic resonance
HRP	Horse radish peroxidase
DBAE	2-(Dibutylamino)ethanol
BCP	Biocatalytic precipitation
CA	Caffeic acid
LOP	Low-oxidation-potential
CRP	C-reactive protein
SCMPs	Streptavidin-coated magnetic particles
AFP	Alpha-fetoprotein

MFPA	Methyl-9-(p-formylphenyl) acridinium carboxylate fluorosulfonate
AZI	Azithromycin
ACE	Acetylspiramycin
ERY	Erythromycin
JOS	Josamycin
IBAN	Ibandronate
ENR	Enrofloxacin
CIP	Ciprofloxacin
GLY	Glyphosate
DLT	Diltiazem hydrochloride
HAS	Human serum albumin
PDDA	Poly(diallyldimethylammonium chloride)
BTPPO	Bis-[3,4,6-trichloro-2-(pentyloxycarbonyl)-phenyl]oxalate
PLANC	Poly(luminol–aniline) nanowires composite
DMBA	4-(Dimethylamino) butyric acid
IgG	Immunoglobulin G
ERGO	Electrochemically reduced graphene oxide
TNTs	Titanate nanotubes
GCE	Glassy carbon electrode
TMB	3,3',5,5'-tetramethylbenzidine
NACE	Nonaqueous capillary electrophoresis
PIN	Positive-intrinsic-negative
PCR	Polymerase Chain Reactions
PVP	Poly(vinylpyridine)
OTA	Ochratoxin A
SELEX	Systematic evolution of ligands by exponential enrichment
FTO	Fluorine-doped tin oxide
TBuA	Tri-n-butylamine
TisoBuA	Tri-isobutylamine
MeDPrA	Methyl-di-n-propylamine
TEtA	Triethylamine
TMeA	Trimethylamine
PCB	Printed circuit board
APD	Avalanche photodiodes
MSD	Mesoscale discovery
SAMs	Self-assembled monolayers
PPV	Polyphenylene vinylene
MP	10-methylphenothiazine

Chapter 1
Introduction

Abstract Chemiluminescence (CL) produced directly or indirectly as a result of electrochemical reactions is known as electrochemiluminescence (ECL), which is from the family of spectro-electrochemical techniques. For better understanding of ECL, various luminescence phenomena, particularly photoluminescence (PL), CL, and ECL, are briefly introduced. A brief ECL research background and theoretic basics of ECL are also described.

Keywords Luminescence · Photoluminescence · Chemiluminescence · Electrochemiluminescence · Triplet–triplet annihilation · Light-emitting reaction · Excimers/exciplexes

Light can be emitted in a number of ways. Luminescence, the generation of light without heat, has diverse types with different origins (Table 1.1). Among this diversity, photoluminescence (PL), chemiluminescence (CL), and electrochemiluminescence (ECL) are the important ones [1]. Let us have a brief idea of PL and CL; ECL will be elaborated in detail.

1.1 Types of Luminescence

1.1.1 Photoluminescence

Photochemical reactions distinctively involve a conversion of light energy to chemical energy (Scheme 1.1) [2].

Scheme 1.1 Conversion of light to chemical energy

$A + B + h\nu \rightarrow$ excited states $\rightarrow$ products
(Light as a reactant $\rightarrow$ conversion. of energy $\rightarrow$ chem. energy)

S. Parveen et al., *Electrogenerated Chemiluminescence*,
SpringerBriefs in Molecular Science, DOI: 10.1007/978-3-642-39555-0_1,
© The Author(s) 2013

Table 1.1 Different types of luminescence[a]

Luminescence type	Caused by	References
Photoluminescence	Photoexcitation of compounds	[4, 5, 63]
Chemiluminescence	Chemical excitation of compounds	[4, 5, 63–75]
Bioluminescence	Luminous organisms	[76]
Electrochemiluminescence	Electrogenerated chemical excitation	[1, 36, 77–89]
Electroluminescence	Radiative recombination of electrons and holes in a material (usually a semiconductor) after an electric current passes through the material or a strong electric field is applied	[3]
Radiochemiluminescence	Radiation-induced chemical excitation	[90–92]
Lyoluminescence	Excitation induced by dissolution of an irradiated or other energy-donating solid	[93]
Sonoluminescence	Excitation of compounds by ultrasonication, either by energy transfer from intrinsic SL centers of water or by chemical excitation by hydroxyl radicals and atomic hydrogen	[94–98]
Pyroluminescence	Flame-excited metal atoms	[99]
Thermoluminescence	Solids subjected to mild heating	[99]

[a] Modified from Refs. [1] and [99]

This light is emitted from transitions between excited and the ground states as indicated in Fig. 1.1. Absorption and emission transitions are involved, such as S_0 to S_1 absorption involves transitions from the ground vibrational state ($v = 0$) of S_0 to various vibrational levels of S_1, ensuing in the observed absorption band. Emission occurs due to transitions from the $v = 0$ state of S_1 to the different vibrational levels of S_0 (Fig. 1.1), resulting in the emission band. Emission bands have lower energies and longer wavelengths in comparison with the absorption band. Stokes shift is a term used for the difference in energy between the absorption and emission bands for transition between the $v = 0$ states (0, 0 bands), which is caused by instantaneous optical transitions ($\sim 10 = 15$ s) where no structural or salvation transformation come about during the transition. During excitation, the transition is from a ground-state structure to an excited-state structure of the same configuration. Though, the excited state can undergo some structural changes (due to dissimilar electronic configuration) to relax to a lower energy state. A transition from this state occurs to the ground state (leading to the emission process) that has the configuration of the excited state. The energy of the singlet transition can be estimated by taking the average of the energies of 0, 0 bands for excitation and emission or from the wavelength λ, where the excitation and emission bands cross, by the formula ES (in eV) $= 1239.81/\lambda$ (in nm). Emission from the triplet state results in phosphorescence that cannot be observed for organic molecules in solution, since these molecules are readily quenched by

Fig. 1.1 Representation of
energy levels and molecular
orbitals during the absorption
and emission of radiation
(Reprinted with permission
from Ref. [3]. Copyright
2004 Taylor & Francis)

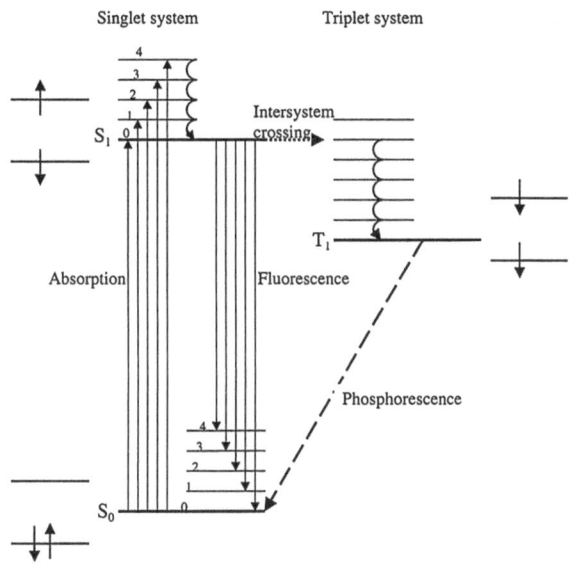

solution species before emission. However, emission from triplet states of metal
chelates is observed, which has much shorter emission lifetimes as compared to
organic molecules. Triplet states have the feature of producing excited singlets, a
type of energy up-conversion, by triplet–triplet annihilation (TTA) (Eq. 1.1).

$$^3A^* + {}^3A^* \rightarrow {}^1A^* A \text{ (triplet--triplet annihilation)} \tag{1.1}$$

Spectroscopic experiments pronounce it as delayed fluorescence. In such an
experiment, a species is first excited by pulse irradiation to the S_1 state, and
emission is delayed until all of the S_1 states have decayed. This decay process may
occur either by radiating to the ground state or by radiationless processes such as
intersystem crossing to the T_1 state. S_1 emission after such a delay results from
transformation of the triplets to excited singlets. Note that the spectroscopy of
delayed fluorescence is not observed for some cases. Though, such systems can be
dealt by the formation of triplets in ECL on account of adjusting the energetics of
annihilation reaction, where TTA is then observed. Thus, states are sometimes
accessible in ECL that is not available spectroscopically [3].

Despite of the need for less expensive optics and light sources, PL will certainly
constantly remain the most significant analytical luminescence method. PL can be
divided into two main subclasses: fluorescence and phosphorescence, and these
analytical methods, including CL, are often reviewed in various articles [4, 5] and
in several other sources [6–8].

1.1.2 Chemiluminescence Reactions

CL is the process of light generation on account of chemical energy. It is most often explained as a dark-field technique where the lack of strong background light level (as found in absorptiometric techniques) diminishes the background signal ensuing in improved detection limits [9].

CL can be well thought-out as the reverse of the Scheme 1.1 of photochemistry (Scheme 1.2) [4].

Scheme 1.2 Conversion of chemical energy to light

$$C + D \rightarrow \text{radical ions} \rightarrow \text{excited states} \rightarrow h\nu \text{ emission}$$

(chem. energy transformation of energy light as a product)

Various compounds react with oxygen or peroxide resulting in the decomposition of specie which in turn generates excited states and emission. For instance, in luminol (5-amino-2,3-dihydro-1,4-phthalazinedione) chemiluminescent reaction (Scheme 1.3), the destruction of luminol takes place with the emitter and the excited phthalate, completely different from the starting material (luminol).

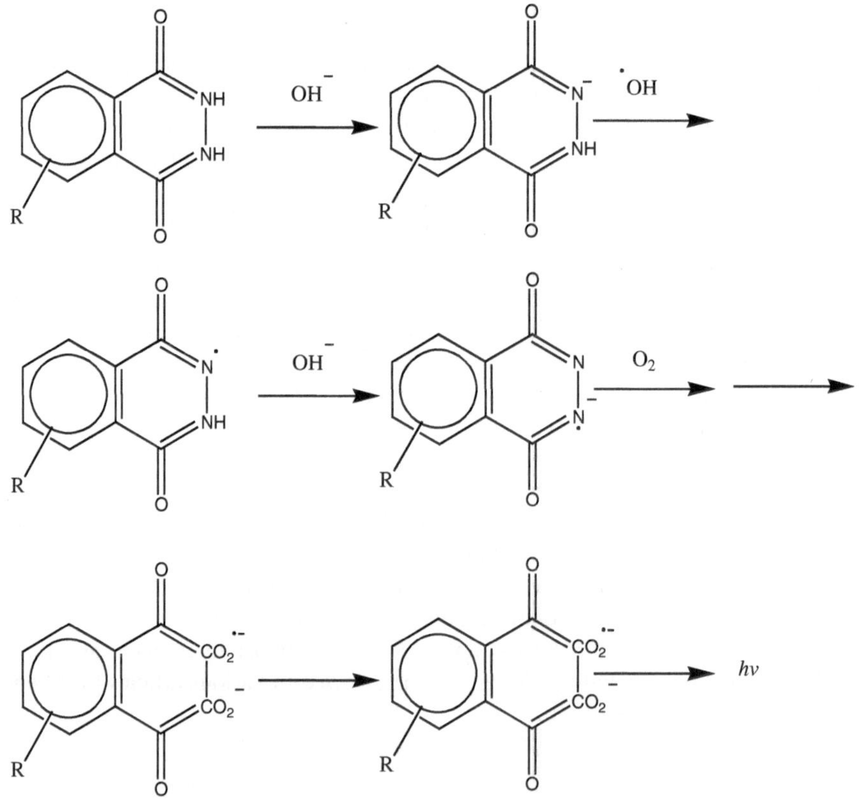

Scheme 1.3 Luminol (5-amino-2,3-dihydro-1,4-phthalazinedione) chemiluminescent reaction

The reaction is triggered by the addition of hydrogen peroxide in the presence of an ion such as Fe^{2+}, which generates hydroxyl radical following a complicated mechanism. CL can be generated in an electrochemical cell by generating peroxide through the reduction of oxygen or oxidation of luminol [3, 10]. For many decades, CL has been recognized as a technique for the analysis of a wide range of compounds both in the liquid and in the gas phase. New strategies based on CL for the sake of improving sensitivity continue to emerge exponentially and have been the focus of comprehensive reviews [11]. CL analysis has a number of advantages including specificity and simplicity, and cheap instrumentation. Moreover, very low limits of detection (LOD) (even down to subfemtomole level) and wide dynamic working range over several orders of magnitude are also observed [12].

Speaking in a broad spectrum, photochemistry is a chemical discipline where light energy transformation occurs into chemical energy and chemical energy into light energy. While electrochemistry can be well thought-out as encompassing both ways of energy; electrical energy transformed into chemical energy and vice versa, sometimes generating highly reactive radical ions. Electrochemically generated luminescence in fact establishes bond between photochemistry and electrochemistry, where chemical reactions of electrochemically generated radicals or radical ions result in excited states which manifest themselves by emitting light [4].

1.1.3 Electro-Chemiluminescence Reactions

Electrogenerated CL (also called ECL and abbreviated as ECL) is the process in accordance with which, species generated at electrodes undergo high-energy electron-transfer reactions to form excited states that emit light (Scheme 1.4) [13].

Scheme 1.4 Conversion of electric energy to light

$$E + F + \text{electricity} \rightarrow \text{rad. ions} \rightarrow \text{excit. states} \rightarrow h\nu \text{ emission}$$

$$(\text{electric energy} \rightarrow \text{chemical energy} \rightarrow \text{light as a product})$$

ECL emission occurs in the visible part of spectrum as a consequence of fast and highly exo-energetic electron-transfer reactions between a strong electron donor and electron acceptor which in turn results in the generation of excited states [4, 14].

ECL is a form of luminescence where the production of light occurs by species that can undergo highly energetic electron-transfer reactions; however, luminescence in CL is initiated by the addition of one necessary reagent to the other.

Luminescence in ECL is initiated and controlled by altering an electrode potential, while in CL, luminescence is often controlled by the careful handling of fluid flow. In contrast, Fig. 1.2 schematically describes the general principles of PL, CL, and ECL. ECL (as an analytical technique) also possesses several advantages over CL [15].

Fig. 1.2 Schematic diagrams showing the general principles of **a** photoluminescence, **b** chemiluminescence, and **c** electrogenerated Chemiluminescence (Reprinted with permission from Ref. [1]. Copyright 2008 American Chemical Society)

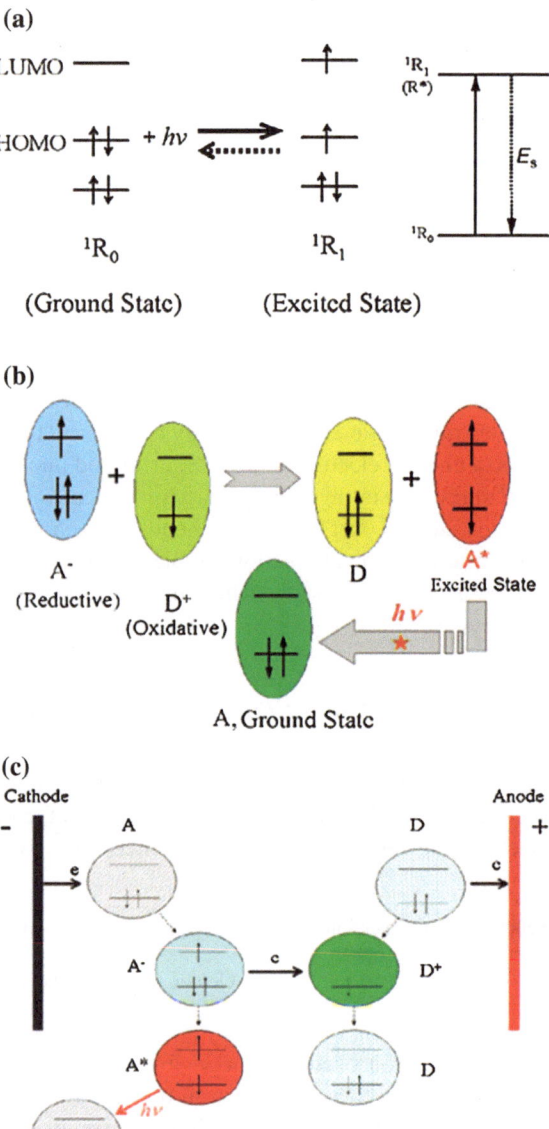

Firstly, in ECL, the electrochemical reaction allows the time and position of the light-emitting reaction to be controlled. (a) Control over time can lead delayed light emission until events take place, such as immune or enzyme-catalyzed reactions have taken place. (b) By controlling position, light emission can be restrained to a region that is precisely located according to the position of detector, increasing the ratio of signal to noise to improve sensitivity, i.e., the combination of ECL with magnetic bead technology, which allows bound label to be distinguished from unbound label without a separation step. Control over position could also be used to determine the results of more than one analytical reaction in the same sample by interrogating each electrode in an array, either in sequence or in simultaneously using a position sensitive detector [16].

Secondly, ECL is more selective than CL, for the reason that the generation of excited states in ECL can be selectively controlled by varying the electrode potentials.

Thirdly, ECL is a nondestructive technique in some cases, because some of the ECL emitters can be regenerated after the ECL emission.

Since ECL is an analytical tool that produces light at an electrode surface; Miao in his excellent review article states that "ECL represents a marriage between electrochemical and spectroscopic methods" [1]. In general, ECL has gained considerable attention as a powerful analytical method owing to its innate features, such as high sensitivity, selectivity, and wide linear dynamic range for a variety of analytes [17–19] including oxalate [20], alkylamines [21], amino acids [22, 23], NADH [24–26], DNA [27], and a number of pharmaceutical compounds [28–31].

ECL is also considered superior over other spectroscopic detection systems due to many distinct advantages [32], such as lack of a light source, the absence of a background optical signal, less sample volume, easy and simple instrumentation, precise control of reaction kinetics offered by controlling the applied potential, compatibility with solution-phase and thin-film formats, and opportunities to enhance intensity with nanomaterials such as metallic nanoparticles (NPs) and nanotubes (NTs), etc [33].

1.2 Research Background of Electrogenerated Chemiluminescence

In the mid of 1960s of the twentieth century, ECL has been studied for the first time on rubrene, 9,10-diphenylanthracene (DPA) and related compounds [34, 35]. Since then so many researchers are thriving for new luminophores, applications, fundamental analysis of its mechanisms, making its instrumentation easier, and achieving lower LOD. In this regard, several review articles, chapters, and monographs have been published to give readers more concise ideas related to ECL [3, 12, 18, 32, 36–53]. Patents on ECL were also applied on July 13, 1964 [54] and August 18, 1964 [55].

Few reports on the subject of light emission during electrolysis were found as early in 1920s [56, 57], while first detailed studies on ECL were addressed in the mid-1960s, by Hercules and Bard [35, 58, 59]. They employed Pt electrodes in acetonitrile (MeCN) or dimethyl formamide (DMF) solutions and presented a detailed study on light emission using DPA, anthracene, and rubrene as emitters. Since the early 1970s, the development of ECL is also connected with the contribution of Prof. Fritz Pragst's scientific work. He benefited significantly to the field of organic mechanisms, electron- or energy transfer reactions, and exploration of triplet states, therefore interrelating organic chemistry, electrochemistry, and photochemistry [4].

Wujian Miao illustrated a time line of various events in the development of ECL till 2002 (Fig. 1.3) [1]. As the time went on, this field attracted bulk of people to do research on ECL basic theory, emitters, mechanisms, applications, etc. Hence, advancements in the area of ECL increased exponentially over more than 45 years. After a long journey of almost half a century, ECL has now grown to be an incredibly potent analytical technique and been extensively used in many areas, such as criminology, forensic, environment, biomedical, biowarfare agent detection immunoassay [3], etc. This technique has also been effectively employed as a detector of flow injection analysis (FIA), high-performance liquid chromatography (HPLC), capillary electrophoresis (CE), and micro total analysis (µTAS) [13].

One of the excellent reviews in the area of ECL, while doing a literature survey using SciFinder Scholar, reveals that more than 2,000 journal articles, book chapters, and patents on various aspects of ECL have been published. For the last few decades, number of publications increased exponentially; many of which were bio-related [1]. In a large part of rest of publications, ECL is also used for analytical applications as a diagnostic tool; however, its contributions to organic chemistry, photo (electro) chemistry, and photophysics are underestimated and ignored [4]. A considerable number of articles on diverse ECL aspects have also been on hand in the literature.

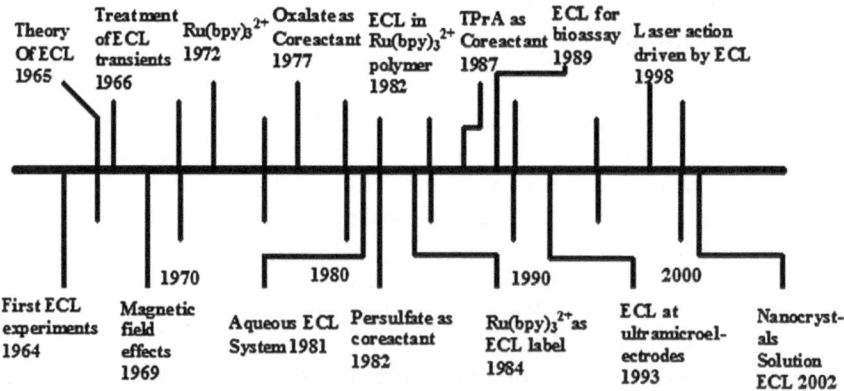

Fig. 1.3 Time line of ECL: 1964–1965, first experiments; 1965, theory 1966, transients; 1969, magnetic field effects; 1972, Ru(bpy)$_3^{2+}$; 1977, oxalate; 1981, aqueous; 1982, Ru(bpy)$_3^{2+}$ polymer and persulfate; 1984, Ru(bpy)$_3^{2+}$ label; 1987, tri-n-propylamine (TPA); 1989, bioassay; 1993, ultramicroelectrodes; 1998, laser action; 2002, semiconductive nanocrystals (Reprinted with permission from Ref. [1]. Copyright 2008 American Chemical Society)

1.3 Basics of Electrogenerated Chemiluminescence

Electrochemical luminescence, a phenomenon where reactants are first electro-chemically generated via two mechanisms: annihilation and coreactants and then combine resulting in the production of light. Both of these mechanisms have various advantages and follow separate pathways to generate the electronically excited state that eventually emits light. Though, all ECL mechanisms involve electrochemically generated chemical species that react and result in emission of light. This method is different from both electroluminescence and CL. The former is a process of electron–hole pair recombination, while the later one in a broader term encircles chemical species that may not be electrochemically generated [1].

Although one of the most important element in the generation of efficient ECL is kinetics of the reaction, however, some other fundamental elements are required for efficient annihilation ECL including (1) radical ion stability of the precursor molecules in the electrolyte of interest, (2) good PL efficiency of a product of the electron-transfer reaction, and (3) sufficient energy in the electron-transfer reaction for the production of excited state [5].

The energy of the electron-transfer reaction, if sufficient, leads to the production of excited states. Since the excitation energy is primarily closest to a thermody-namic internal energy, the excited singlet state of energy ES (in electronvolts) can be produced as shown in Eq. 1.2

$$-\Delta H^0 = E^\circ \, (D^{+\bullet}, D) - E^\circ (A, A^{-\bullet}) - T\Delta S^0 > \text{ES} \qquad (1.2)$$

The value of $T\Delta S^0$ is usually estimated as 0.1 (± 0.1) eV [60–62]. Frequently, the criterion is given based on CV peak potentials (in volts) at $T = 298$ K as Eq. 1.3 [60]

$$E_p(D^{+\bullet}, D) - E_p(A, A^{-\bullet}) - 0.16 > \text{ES} \qquad (1.3)$$

Such reactions are occasionally considered as "energy sufficient," and the reaction is supposed to follow the S route. $-\Delta H^0 > \text{ET}$ shows that the reaction is following T route by triplet state production at energy ET with excited singlet production occurring through TTA. Both these excited singlets and triplets are sometimes produced in significant amounts in the annihilation process following the ST route. The reactions forming excimers or exciplexes (Eqs. 1.4 and 1.5) are thought to proceed by the E route [5].

$$A^{-\bullet}A^\bullet \rightarrow A_2^* \qquad \text{(excimer formation)} \qquad (1.4)$$

$$A^{-\bullet}D^\bullet \rightarrow AD^* \qquad \text{(exciplex formation)} \qquad (1.5)$$

References

1. Miao W (2008) Electrogenerated chemiluminescence and its biorelated applications. Chem Rev 108(7):2506–2553. doi:10.1021/cr068083a
2. Ludvik J (2011) DC-electrochemiluminescence (ECL with a coreactant)-principle and applications in organic chemistry. J Solid State Electrochem 15(10):2065–2081. doi:10.1007/s10008-011-1546-x
3. Bard AJ (2004) Electrogenerated chemiluminescence. Taylor & Francis
4. Agbaria RA, Oldham PB, McCarroll M, McGown LB, Warner IM (2002) Molecular fluorescence, phosphorescence, and chemiluminescence spectrometry. Anal Chem 74(16):3952–3962. doi:10.1021/ac020299z
5. Oldham PB, McCarroll ME, McGown LB, Warner IM (2000) Molecular fluorescence, phosphorescence, and chemiluminescence spectrometry. Anal Chem 72(12):197–210. doi:10.1021/a1000017p
6. Kraayenhof R, Visser AJWG, Gerritsen HC (2002) Fluorescence spectroscopy, imaging, and probes: new tools in chemical, physical, and life sciences. Springer, Berlin
7. Lakowicz JR, Technology CoAiFS (2001) Advances in fluorescence sensing technology V: Fifth SPIE conference on advances in fluorescence sensing technology held as part of photonics West, 24–25 January 2001, San Jose, USA. SPIE, the international society for optical engineering
8. Holden M, Wang L (2008) Quantitative real-time pcr: fluorescent probe options and issues. In: Resch-Genger U (ed) Standardization and quality assurance in fluorescence measurements II, vol 6. Springer series on fluorescence, springer, Berlin Heidelberg, pp 489–508. doi:10.1007/4243_2008_046
9. Gamiz-Gracia L, Garcia-Campana AM, Huertas-Perez JF, Lara FJ (2009) Chemiluminescence detection in liquid chromatography: applications to clinical, pharmaceutical, environmental and food analysis–a review. Anal Chim Acta 640(1–2):7–28. doi:10.1016/j.aca.2009.03.017
10. Kuwana T, Epstein B, Seo ET (1963) Electrochemical generation of solution luminescence. J Phy Chem 67:2243–2244
11. Bowie AR, Sanders MG, Worsfold PJ (1996) Analytical applications of liquid phase chemiluminescence reactions–a review. J Biolumin Chemilumin 11(2):61–90. doi:10.1002/(sici)1099-1271(199603)11
12. Knight AW (1999) A review of recent trends in analytical applications of electrogenerated chemiluminescence. Trac-Trends Anal Chem 18(1):47–62. doi:10.1016/s0165-9936(98)00086-7
13. Yin XB, Dong SJ, Wang E (2004) Analytical applications of the electrochemiluminescence of tris (2.2′-bipyridyl) ruthenium and its derivatives. Trac-Trends in Anal Chem 23(6):432–441. doi:10.1016/s0165-9936(04)00603-x
14. Wei H, Wang E (2011) Electrochemiluminescence of tris(2,2′-bipyridyl)ruthenium and its applications in bioanalysis: a review. Lumin 26(2):77–85. doi:10.1002/bio.1279
15. Wilson R, Clavering C, Hutchinson A (2003) Electrochemiluminescence enzyme immunoassays for tnt and pentaerythritol tetranitrate. Anal Chem 75(16):4244–4249. doi:10.1021/ac034163s
16. Chen X-m, Su B-y, Song X-h, Chen Q-a, Chen X, Wang X-r (2011) Recent advances in electrochemiluminescent enzyme biosensors. Trac-Trends Anal Chem 30(5):665–676. doi:10.1016/j.trac.2010.12.004
17. Knight AW, Greenway GM (1996) Relationship between structural attributes and observed electrogenerated chemiluminescence (ECL) activity of tertiary amines as potential analytes for the tris(2,2-bipyridine)ruthenium(II) ECL reaction: a review. Analyst 121(11):101R–106R
18. Lee WY (1997) Tris (2,2′-bipyridyl)ruthenium(II) electrogenerated chemiluminescence in analytical science. Mikrochim Acta 127(1–2):19–39. doi:10.1007/bf01243160

19. Gerardi RD, Barnett NW, Lewis SW (1999) Analytical applications of tris(2,2′-bipyridyl)ruthenium(III) as a chemiluminescent reagent. Anal Chim Acta 378(1–3):1–41. doi:10.1016/s0003-2670(98)00545-5

20. Rubinstein I, Martin CR, Bard AJ (1983) Electrogenerated chemiluminescent determination of oxalate. Anal Chem 55(9):1580–1582. doi:10.1021/ac00260a030

21. Noffsinger JB, Danielson ND (1987) Generation of chemiluminescence upon reaction of aliphatic amines with tris(2,2′-bipyridine)ruthenium(III). Anal Chem 59(6):865–868. doi:10.1021/ac00133a017

22. Brune SN, Bobbitt DR (1991) Effect of pH on the reaction of tris(2,2′-bipyridyl)ruthenium(III) with amino-acids: implications for their detection. Talanta 38(4):419–424. doi:10.1016/0039-9140(91)80080-J

23. Brune SN, Bobbitt DR (1992) Role of electron-donating/withdrawing character, pH, and stoichiometry on the chemiluminescent reaction of tris(2,2′-bipyridyl)ruthenium(III) with amino acids. Anal Chem 64(2):166–170

24. Downey TM, Nieman TA (1992) Chemiluminescence detection using regenerable tris(2,2′-bipyridyl)ruthenium(II) immobilized in Nafion. Anal Chem 64(3):261–268. doi:10.1021/ac00027a005

25. Lee W-Y, Nieman TA (1995) Evaluation of use of tris(2,2′-bipyridyl)ruthenium(III) as a chemiluminescent reagent for quantitation in flowing streams. Anal Chem 67(11):1789–1796. doi:10.1021/ac00107a007

26. Martin AF, Nieman TA (1993) Glucose quantitation using an immobilized glucose dehydrogenase enzyme reactor and a tris(2,2′-bipyridyl) ruthenium(II) chemiluminescent sensor. Anal Chim Acta 281(3):475–481. doi:10.1016/0003-2670(93)85005-5

27. Dennany L, Forster RJ, Rusling JF (2003) Simultaneous direct electrochemiluminescence and catalytic voltammetry detection of dna in ultrathin films. J Am Chem Soc 125(17):5213–5218. doi:10.1021/ja0296529

28. Zorzi M, Pastore P, Magno F (2000) a single calibration graph for the direct determination of ascorbic and dehydroascorbic acids by electrogenerated luminescence based on Ru(bpy)$_3^{2+}$ in aqueous solution. Anal Chem 72(20):4934–4939. doi:10.1021/ac991222m

29. Greenway GM, Nelstrop LJ, Port SN (2000) Tris(2,2-bipyridyl)ruthenium (II) chemiluminescence in a microflow injection system for codeine determination. Anal Chim Acta 405(1–2):43–50. doi:10.1016/S0003-2670(99)00691-1

30. Park Y-J, Lee DW, Lee W-Y (2002) Determination of β-blockers in pharmaceutical preparations and human urine by high-performance liquid chromatography with tris(2,2′-bipyridyl)ruthenium(II) electrogenerated chemiluminescence detection. Anal Chim Acta 471(1):51–59. doi:10.1016/S0003-2670(02)00932-7

31. Li F, Cui H, Lin X-Q (2002) Determination of adrenaline by using inhibited Ru(bpy)$_3^{2+}$ electrochemiluminescence. Anal Chim Acta 471(2):187–194. doi:10.1016/S0003-2670(02)00930-3

32. Bard AJ, Debad JD, Leland JK, Sigal GB, Wilbur JL, Wohlsatdter JN (2000) Encyclopedia of analytical chemistry: applications, theory and instrumentation, vol 11. Wiley, New York

33. Forster RJ, Bertoncello P, Keyes TE (2009) Electrogenerated Chemiluminescence. Ann Rev Anal Chem 2:359–385. doi:10.1146/annurev-anchem-060908-155305

34. Swanick KN, Dodd DW, Price JT, Brazeau AL, Jones ND, Hudson RHE, Ding Z (2011) Electrogenerated chemiluminescence of triazole-modified deoxycytidine analogues in N-dimethylformamide. Phy Chem Chem Phy 13(38):17405–17412

35. Santhanam KSV, Bard AJ (1965) J Am Chem Soc 87:139–140

36. Hu L, Xu G (2010) Applications and trends in electrochemiluminescence. Chem Soc Rev 39(8):3275–3304. doi:10.1039/b923679c

37. Bard AJ, Rubinstein I (2003) Electroanalytical Chemistry: A series of advances. vol 22; v. 2004. Taylor & Francis

38. Adam W, Cilento G (1982) Chemical and biological generation of excited states. Academic Press

39. Aikens DA (1983) Electrochemical methods, fundamentals and applications. J Chem Edu 60(1):A25. doi:10.1021/ed060pA25.1

40. Knight AW, Greenway GM (1994) Occurrence, mechanisms and analytical applications of electrogenerated chemiluminescence: review. Analyst 119(5):879–890. doi:10.1039/an9941900879

41. Gerardi RD, Barnett NW, Lewis SW (1999) Analytical applications of tris(2,2′-bipyridyl)ruthenium(III) as a chemiluminescent reagent. Anal Chim Acta 378(1–3):1–41. doi:10.1016/S0003-2670(98)00545-5

42. Kukoba AV, Bykh AI, Svir IB (2000) Analytical applications of electrochemiluminescence: an overview. Fresenius J Anal Chem 368(5):439–442. doi:10.1007/s002160000548

43. Mitschke U, Bauerle P (2000) The electroluminescence of organic materials. J Mat Chem 10(7):1471–1507

44. Garcia-Campana AM (2001) Chemiluminescence in analytical chemistry. Taylor & Francis

45. Isacsson U, Wettermark G (1974) Chemiluminescence in analytical chemistry. Anal Chim Acta 68(2):339–362. doi:10.1016/S0003-2670(01)82590-3

46. Andersson AM, Schmehl RH (2001) Molecular and supramolecular photochemistry 7: Optical Sensors and Switches

47. Fahnrich KA, Pravda M, Guilbault GG (2001) Recent applications of electrogenerated chemiluminescence in chemical analysis. Talanta 54(4):531–559. doi:10.1016/s0039-9140(01)00312-5

48. Ligler FS, Taitt CR (2002) Optical biosensors: present and future. Elsevier Science

49. Kulmala S, Suomi J (2003) Current status of modern analytical luminescence methods. Anal Chim Acta 500(1–2):21–69. doi:10.1016/j.aca.2003.09.004

50. Knight AW (1999) A review of recent trends in analytical applications of electrogenerated chemiluminescence. TrAC, Trends Anal Chem 18(1):47–62. doi:10.1016/S0165-9936(98)00086-7

51. Richter MM (2004) Electrochemiluminescence (ECL). Chem Rev 104(6):3003–3036. doi:10.1021/cr020373d

52. Gorman BA, Francis PS, Barnett NW (2006) Tris(2,2′-bipyridyl)ruthenium(II) chemiluminescence. Analyst 131(5):616–639. doi:10.1039/b518454a

53. Pyati R, Richter MM (2007) ECL-Electrochemical luminescence. Ann Rep Sec "C" (Phy Chem) 103(0):12–78

54. Maricle DL, Rauhut MM (1972) US Patent

55. Chandross EA, Visco RE (1967) 3,319,132, May 9

56. Dufford RT, Nightingale D, Gaddum LW (1927) Luminescence of grignard compounds in electric and magnetic fields, and related electrical phenomena. J Am Chem Soc 49(8):1858–1864. doi:10.1021/ja01407a002

57. Harvey N (1928) Luminescence during electrolysis. J Phy Chem 33(10):1456–1459. doi:10.1021/j150304a002

58. Hercules DM (1964) Chemiluminescence resulting from electrochemically generated species. Science (New York, NY) 145 (3634):808–809. doi:10.1126/science.145.3634.808

59. Visco RE, Chandross EA (1964) Electroluminescence in solutions of aromatic hydrocarbons. J Am Chem Soc 86(23):5350–5351

60. Faulkner LR, Bard AJ Electroanalytical Chemistry, vol 10. Marcel Dekker, New York

61. Weller A, Zachariasse K (1967) Chemiluminescence from chemical oxidation of aromatic anions. J Chem Phy 46(12):4984–4985

62. Faulkner LR, Tachikawa H, Bard AJ (1972) Electrogenerated chemiluminescence. VII. Influence of an external magnetic field on luminescence intensity. J Am Chem Soc 94(3):691–699. doi:10.1021/ja00758a001

63. Powe AM, Das S, Lowry M, El-Zahab B, Fakayode SO, Geng ML, Baker GA, Wang L, McCarroll ME, Patonay G, Li M, Aljarrah M, Neal S, Warner IM (2010) Molecular fluorescence, phosphorescence, and chemiluminescence spectrometry. Anal Chem 82(12):4865–4894. doi:10.1021/ac101131p

64. Aitken RJ, Baker MA, O'Bryan M (2004) Andrology lab corner: shedding light on chemiluminescence: the application of chemiluminescence in diagnostic andrology. J Andr 25(4):455–465. doi:10.1002/j.1939-4640.2004.tb02815.x

65. Butkovskaya NI, Setser DW (2003) Infrared chemiluminescence from water-forming reactions: characterization of dynamics and mechanisms. Int Rev Phy Chem 22(1):1–72. doi:10.1080/0144235021000033381

66. Gaffney JS, Anl A IL, Marley NA (2002) Historical overview of the development of chemiluminescence detection and its application to air pollutants. In: Atmospheric chemistry: urban, regional and global-scale impacts of air pollutants, Boston, MA. American Meteorological Society, pp 1–6

67. Garcia-Campana A, Ugent WB, Cuadros-Rodriguez L, Barrero F, Bosque-Sendra J, Gamiz-Gracia L (2002) Curr Org Chem 6(1):1–20

68. Jacobson K, Eriksson P, Reitberger T, Stenberg B (2004) Chemiluminescence as a tool for polyolefin oxidation studies. In: Albertsson A-C (ed) Long Term Properties of Polyolefins, vol 169. Advances in Polymer Science. Springer Berlin Heidelberg, pp 151–176. doi:10.1007/b13522

69. Kuyper C, Milofsky R (2001) Recent developments in chemiluminescence and photochemical reaction detection for capillary electrophoresis. TrAC Trends Anal Chem 20(5):232–240. doi:10.1016/S0165-9936(01)00066-8

70. Zhan W, Alvarez J, Sun L, Crooks RM (2003) A multichannel microfluidic sensor that detects anodic redox reactions indirectly using anodic electrogenerated chemiluminescence. Anal Chem 75(6):1233–1238

71. Liu Y-M, Cheng J-K (2002) Ultrasensitive chemiluminescence detection in capillary electrophoresis. J Chromatogr A 959(1–2):1–13. doi:10.1016/S0021-9673(02)00434-X

72. Roda A, Guardigli M, Michelini E, Mirasoli M, Pasini P (2003) Peer reviewed: analytical bioluminescence and chemiluminescence. Anal Chem 75 (21):462 A–470 A. doi:10.1021/ac031398v

73. Roda A, Guardigli M, Pasini P, Mirasoli M (2003) Bioluminescence and chemiluminescence in drug screening. Anal Bioanal Chem 377(5):826–833. doi:10.1007/s00216-003-2096-6

74. Stott RA (2002) Enhanced chemiluminescence immunoassay, pp 1089–1096

75. Yamaguchi M, Yoshida H, Nohta H (2002) Luminol-type chemiluminescence derivatization reagents for liquid chromatography and capillary electrophoresis. J Chromatogr A 950(1–2):1–19

76. Harvey EN (1952) Bioluminescence. Academic Press

77. Qian J, Zhang C, Cao X, Liu S (2010) Versatile immunosensor using a quantum dot coated silica nanosphere as a label for signal amplification. Anal Chem 82(15):6422–6429. doi:10.1021/ac100558t

78. Wei H, Liu J, Zhou L, Li J, Jiang X, Kang J, Yang X, Dong S, Wang E (2008) Ru(bpy)$_3^{2+}$-doped silica nanoparticles within layer-by-layer biomolecular coatings and their application as a biocompatible electrochemiluminescent tag material. Chem Eur J 14(12):3687–3693. doi:10.1002/chem.200701518

79. Xu XHN, Jeffers RB, Gao JS, Logan B (2001) Novel solution-phase immunoassays for molecular analysis of tumor markers. Analyst 126(8):1285–1292. doi:10.1039/b104180k

80. Yuan Y, Li H, Han S, Hu L, Parveen S, Xu G (2011) Vitamin C derivatives as new coreactants for tris(2,2′-bipyridine)ruthenium(II) electrochemiluminescence. Anal Chim Acta 701(2):169–173. doi:10.1016/j.aca.2011.06.051

81. Arai K, Takahashi K, Kusu F (1999) An electrochemiluminescence flow through-cell and its applications to sensitive immunoassay using N-(aminobutyl)-N-ethylisoluminol. Anal Chem 71(11):2237–2240. doi:10.1021/ac9810361

82. Jie G, Liu P, Wang L, Zhang S (2010) Electrochemiluminescence immunosensor based on nanocomposite film of CdS quantum dots-carbon nanotubes combined with gold nanoparticles-chitosan. Electrochem Comm 12(1):22–26. doi:10.1016/j.elecom.2009.10.027

83. Parveen S, Zhang W, Yuan Y, Hu L, Shah Gilani MRH, Rehman Au, Xu G (2013) Electrogenerated chemiluminescence of/2-(dibutylamino) ethanol system. J Electroanal Chem 688:45–48. doi:10.1016/j.jelechem.2012.05.014

84. Marquette CA, Blum LJ (1998) Electrochemiluminescence of luminol for 2,4-D optical immunosensing in a flow injection analysis system. Sens Actua B-Chem 51(1–3):100–106. doi:10.1016/s0925-4005(98)00175-0

85. Yuan Y, Han S, Hu L, Parveen S, Xu G (2012) Coreactants of tris(2,2'-bipyridyl)ruthenium(II) Electrogenerated Chemiluminescence. Electrochim Acta 82:484–492. doi:10.1016/j.electacta.2012.03.156

86. Sun S, Yang M, Kostov Y, Rasooly A (2010) ELISA-LOC: lab-on-a-chip for enzyme-linked immunodetection. Lab Chip 10(16):2093–2100. doi:10.1039/c003994b

87. Han S, Zhu S, Liu Z, Hu L, Parveen S, Xu G (2012) Oligonucleotide-stabilized fluorescent silver nanoclusters for turn-on detection of melamine. Biosens Bioelectron 36(1):267–270. doi:10.1016/j.bios.2012.04.028

88. Han B, Du Y, Wang E (2008) Simultaneous determination of pethidine and methadone by capillary electrophoresis with electrochemiluminescence detection of tris(2,2'-bipyridyl)ruthenium(II). Microchem J 89(2):137–141. doi:10.1016/j.microc.2008.01.007

89. Yuan Y, Li H, Han S, Hu L, Parveen S, Cai H, Xu G (2012) Immobilization of tris(1,10-phenanthroline)ruthenium with graphene oxide for electrochemiluminescent analysis. Anal Chim Acta 720:38–42. doi:10.1016/j.aca.2012.01.023

90. Bistolfi F (2000) Red radioluminescence and radiochemiluminescence: premises for a photodynamic tumour therapy with X-rays and haematoporphyrin derivatives. A working hypothesis. Panminerva medica 42(1):69–75

91. Papadopoulos K, Lignos J, Stamatakis M, Dimotikali D, Nikokavouras J (1998) Radiochemiluminescence of acridones and alkyl acridities. J Photochem Photobio A: Chem 115(2):137–142. doi:10.1016/S1010-6030(98)00236-6

92. Papadopoulos K, Triantis T, Dimotikali D, Nikokavouras J (2000) Radiochemiluminescence of carboxyquinolines. J Photochem Photobio A: Chem 131(1–3):55–60. doi:10.1016/S1010-6030(99)00243-9

93. Håkansson M, Jiang Q, Spehar A-M, Suomi J, Kulmala S (2006) Extrinsic lyoluminescence of aluminum induced by lanthanide chelates in alkaline aqueous solution. J Lumin 118(2):272–282. doi:10.1016/j.jlumin.2005.09.012

94. Ashokkumar M, Grieser F (2004) Single bubble sonoluminescence–a chemist's overview. Chemphyschem: Eur j chem phy phy chem 5 (4):439–448. doi:10.1002/cphc.200300926

95. Arakeri VH (2003) Sonoluminescence and bubble fusion. Curr Sci 85(7):911–916

96. Brenner MP, Hilgenfeldt S, Lohse D (2002) Single-bubble sonoluminescence. Rev Mod Phy 74(2):425–484

97. Hammer D, Frommhold L (2001) Sonoluminescence: how bubbles glow. J Mod Opt 48(2):239–277. doi:10.1080/09500340108232457

98. Prevenslik TV (2003) The cavitation induced Becquerel effect and the hot spot theory of sonoluminescence. Ultrason 41(4):313–317. doi:10.1016/S0041-624X(02)00458-4

99. Marquette C, Blum L (2008) Electro-chemiluminescent biosensing. Anal Bioanal Chem 390(1):155–168. doi:10.1007/s00216-007-1631-2

Chapter 2
Generation Pathways of Electrogenerated Chemiluminescence

Abstract ECL continues to be an area of active research. This chapter provides a brief way for understanding fundamentals of ECL. An overview of selected key ECL mechanisms for the production of ECL is given. Studies on finding new ECL co-reactants, disclosing the relationship between ECL efficiencies and structure of co-reactants, and improving ECL efficiencies are also discussed.

Keywords Annihilation · Pathway · Electron-transfer reaction · Gibbs free energy · Energy sufficient reaction · Reductive–oxidation co-reactant · Oxidative–reduction co-reactant · Hot-electron ECL

Electrogenerated chemiluminescence (ECL) can be generated by two principal routes, namely the annihilation and co-reactant pathways. In each case, two species are generated electrochemically, and those two species undergo an electron-transfer reaction to produce an emissive species [1].

2.1 Annihilation ECL Pathway

The early ECL studies originated with ion annihilation ECL [2]. Proceeding by annihilation pathway, ECL can be generated on reaction between oxidized and reduced species produced either at a single electrode by using an alternating potential or at two separate electrodes. In later case, one electrode is set at a reductive potential and the other at an oxidative potential and they are placed in close proximity to each other [3]. Most of the annihilation ECL processes have been investigated in organic solvents or partially organic solutions because annihilation reactions are very energetic and aqueous solutions usually have too narrow potential range to allow convenient electrolytic generation of both the oxidized and reduced ECL precursors [4]. Such reactions performed in organic solutions require a supporting electrolyte such as tetrabutylammonium salt and the absence of dissolved oxygen as this oxygen may sometimes react with radical intermediates and quenches the reaction.

S. Parveen et al., *Electrogenerated Chemiluminescence*,
SpringerBriefs in Molecular Science, DOI: 10.1007/978-3-642-39555-0_2,

15

A system having radical cations and anions from different molecules may also exhibit annihilation processes. Gibbs free energy for annihilation process can be calculated from the redox potentials using Eq. (2.1).

$$\Delta G = -nF\left(E^{\circ}{}_{reduction} - E^{\circ}{}_{oxidation}\right) \tag{2.1}$$

where ΔG is Gibbs free energy, F is Faraday's constant, while $E^{\circ}{}_{reduction} - E^{\circ}{}_{oxidation}$ are the reduction and oxidation potentials, respectively. As enthalpy is directly related to Gibbs free energy, enthalpy can be calculated based on Eq. (2.2).

$$\Delta G = \Delta H - T\Delta S \tag{2.2}$$

Enthalpy plays a significant role in describing which path to be followed, i.e.,

1. The enthalpy exceeds the energy required to produce the lowest excited states from the ground state. In this case, the reaction is to follow the singlet route "S-route" and 1R* will be directly generated. The reaction is recognized as "energy sufficient," and DPA system is the notorious example of this case.
2. The enthalpy is lower than the energy required to produce the lowest excited state, still exceeding the triplet state energy, 3R*, which in turn produces 1R* by subsequent annihilation of 3R* (triplet–triplet annihilation, TTA). ECL of ruthenium tris-bipyridyl-type derivatives falls in this category. Additionally, ion annihilation can also follow "E-route," ensuing the formation of excimers (excited dimers) and exciplexes (excited complexes).

The key advantage of the annihilation process is its simplistic approach as it requires only the ECL species, solvent, and supporting electrolyte in order to generate light [5]. The mechanism of annihilation method is generalized as shown in Scheme 2.1, where A is a polyaromatic hydrocarbon (PAH), B is either the same or another PAH or an aromatic derivative, 1A* is the singlet excited state, and 3A* is the triplet excited state [6].

Let us have a common example of the above annihilation method. First of all, Bard's group [7] in 1972 published an article related to the ECL of $Ru(bpy)_3{}^{2+}$

Scheme 2.1 Mechanism of annihilation method [6]

$A + e^- \rightarrow A^{\circ-}$	Electro-reduction
$B - e^- \rightarrow B^{+\circ}$	Electro-oxidation
either	
$A^{\circ-} + B^{+\circ} \rightarrow {}^1A^* + B$	Electron transfer
or	(Energy sufficient system)
$A^{\circ-} + B^{+\circ} \rightarrow {}^3A^* + B$	Electron transfer
${}^3A^* + {}^3A^* \rightarrow {}^1A^* + A$	Triplet-triplet annihilation
followed by	(Energy deficient system)
${}^1A^* \rightarrow A + h\nu$	

Fig. 2.1 Structure of Ru(bpy)$_3^{2+}$ and proposed mechanism for Ru(bpy)$_3^{3+}$/Ru(bpy)$_3^+$ ECL system (Reprinted with permission from Ref. [34]. Copyright 2004 American Chemical Society)

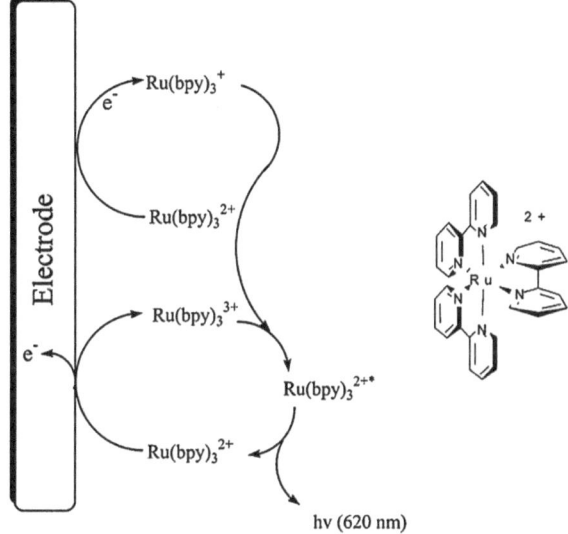

(Fig. 2.1). They generated ECL by alternatively pulsing the electrode potential to form oxidative Ru(bpy)$_3^{3+}$ and reductive Ru(bpy)$_3^+$ in acetonitrile (Eqs. 2.3–2.6) [8].

$$Ru(bpy)_3^{2+} + e^- \rightarrow Ru(bpy)_3^+ \tag{2.3}$$

$$Ru(bpy)_3^{2+} - e^- \rightarrow Ru(bpy)_3^{3+} \tag{2.4}$$

$$Ru(bpy)_3^+ + Ru(bpy)_3^{3+} \rightarrow Ru(bpy)_3^{2+} + Ru(bpy)_3^{2+*} \tag{2.5}$$

$$Ru(bpy)_3^{2+} \rightarrow Ru(bpy)_3^{2+} + hv \tag{2.6}$$

As Ru(bpy)$_3^+$ is exceptionally unstable in aqueous solutions, so it was found that annihilation ECL cannot be observed in aqueous solution by changing the potential successively to generate Ru(bpy)$_3^{3+}$ and Ru(bpy)$_3^+$. However, annihilation ECL of Ru(bpy)$_3^{2+}$ can more likely be attained in aqueous solution employing a carbon-interdigitated microelectrode array of 2 μm width and spacing [9]. These electrodes are kept so close that ECL can be obviously seen with the unaided eye in normal room lighting. The experimental concentration is kept greater than 1 mM in a generation/collection biasing mode. Later, this system was applied by Michel and group for analysis of protein [10].

2.2 Co-reactant ECL Pathway

Currently, overwhelming majority of commercially available ECL analytical instruments is based on co-reactant ECL technology [2]. The reason for its becoming popular and easy availability lies in the key advantage of the co-reactant approach which includes its assistance in ECL generation in aqueous solution, opening a gateway to a wide range of assays for molecules of diagnostic or biological significance. Moreover, co-reactant ECL requires one-directional potential scanning at an electrode in a solution having luminophore species and a reagent (co-reactant), whereas electrolytic generation of both the oxidized and reduced ECL precursors is required in ion annihilation ECL [4, 11]. In ion annihilation ECL, all starting species can be regenerated after light emission, whereas in a co-reactant ECL system, only luminophore species can be regenerated, whereas the co-reactant is consumed via electrochemical–chemical reactions. In situ-generated co-reactant intermediates, as indicated by their standard redox potentials, are either strong reducing agents (in oxidative–reduction ECL) or strong oxidizing agents (in reductive–oxidation ECL) [2]. Noteworthily, co-reactant is consumed during the chemical (or electrochemical) reactions, while only the luminophore species can be regenerated at the electrode. Suitable co-reactants can be easily oxidized or reduced, which undergo a rapid chemical reaction to form an intermediate species with sufficient oxidizing or reducing power to create the excited state of the luminophore [5]. Therefore, to select a good ECL co-reactant, many factors should be kept in mind [12], including solubility, stability, electrochemical properties, kinetics, quenching effect, ECL background. Of these, electrochemical properties of the co-reactant are foremost important factor. The co-reactant should be available for easy oxidation or reduction with the luminophore species at or near the electrode and undergo a rapid chemical reaction to form an intermediate possessing sufficient reducing or oxidizing energy to react with the oxidized or reduced luminophore to form the excited state.

Four processes are usually involved in co-reactant ECL systems (as presented in Table 2.1) such as (a) redox reactions at electrode, (b) homogeneous chemical reactions, (c) excited-state species formation, and (d) light emission. Two types of redox reactions, namely heterogeneous and homogeneous redox reactions of co-reactants, are possible, which depend on the redox potential of the co-reactant and nature of the working electrode [2].

Both, the luminophore and the co-reactant, in co-reactant pathway, can be first oxidized or reduced at the electrode to form radicals by considering the polarity of the applied potential. The intermediates thus formed from the co-reactant then decompose to produce a strong reducing or oxidizing species that reacts with the oxidized or reduced luminophore to produce the excited states that emit light. The term "oxidative–reduction" ECL is usually used when highly reducing intermediate species are generated after an electrochemical oxidation of a co-reactant, and "reductive–oxidation" ECL is the phenomenon where highly oxidizing intermediates are produced after an electrochemical reduction [5].

Table 2.1 General mechanisms of co-reactant ECL systems[a]

Reaction process	Oxidative–reduction ECL	Reductive–oxidation ECL
Redox reactions at electrode	$R - e \rightarrow R^{\bullet+}$ $C - e \rightarrow C^{\bullet+}$	$R + e \rightarrow R^{\bullet-}$ $C + e \rightarrow C^{\bullet-}$
Homogeneous chemical reactions	$R^{\bullet+} + C \rightarrow R + C^{\bullet+}$ $C^{\bullet+} \rightarrow C^{\neq}_{Red}$ $C^{\neq}_{Red} + R \rightarrow R^{\bullet-} + P$	$R^{\bullet-} + C \rightarrow R + C^{\bullet-}$ $C^{\bullet-} \rightarrow C^{\neq}_{Ox}$ $C^{\neq}_{Ox} + R \rightarrow R^{\bullet+} + P$
Excited-state species formation	$R^{\bullet+} + R^{\bullet-} \rightarrow R + R^*$ or $R^{\bullet+} + C^{\neq}_{Red} \rightarrow R^* + P$	$R^{\bullet+} + R^{\bullet-} \rightarrow R + R^*$ or $R^{\bullet-} + C^{\neq}_{Ox} \rightarrow R^* + P$
Light emission	$R^* \rightarrow R + h\nu$	$R^* \rightarrow R + h\nu$

[a] R luminophore, C co-reactant, $C^{\neq}$ co-reactant intermediate with subscript "Red" for reducing agent and "Ox" for oxidizing agent, P product associated with $C^{\neq}$ reactions

2.2.1 "Reductive–Oxidation" Co-reactants

2.2.1.1 Peroxydisulfate (Persulfate, $S_2O_8^{2-}$) System

ECL in the "reductive–oxidation" systems is usually produced by applying a very negative potential. Consequently, it seems quite difficult to study stable ECL in these systems in aqueous solutions because of serious hydrogen evolution upon applying very negative potentials [13]. Persulfate produces ECL on reaction with $Ru(bpy)_3^+$ in DMF or acetonitrile–H_2O mixed solution (Eqs. 2.7–2.12) [10, 14]. It inhibits ECL emission in both solutions, if present in high concentrations. In former solution, inhibition resulted due to side reactions rather from quenching reaction in DMF, while quenching reaction occurred in the acetonitrile–H_2O mixed solution. The ECL emission was not observed in purely aqueous solutions due to instability of $Ru(bpy)_3^+$ or due to the rapid quenching by persulfate in aqueous solution which may be the result of serious hydrogen evolution upon applying very negative potentials.

$$Ru(bpy)_3^{2+} + e^- \rightarrow Ru(bpy)_3^+ \qquad (2.7)$$

$$Ru(bpy)_3^+ + S_2O_8^{2-} \rightarrow Ru(bpy)_3^{2+} + SO_4^{\bullet-} + SO_4^{2-} \qquad (2.8)$$

$$Ru(bpy)_3^+ + SO_4^{\bullet-} \rightarrow Ru(bpy)_3^{2+*} + SO_4^{2-} \qquad (2.9)$$

$$Ru(bpy)_3^{2+} + SO_4^{\bullet-} \rightarrow Ru(bpy)_3^{3+} + SO_4^{2-} \qquad (2.10)$$

$$Ru(bpy)_3^{3+} + Ru(bpy)_3^+ \rightarrow Ru(bpy)_3^{2+*} + Ru(bpy)_3^{2+} \qquad (2.11)$$

$$Ru(bpy)_3^{2+*} \rightarrow Ru(bpy)_3^{2+} + h\nu \qquad (2.12)$$

In 1984, Bard first of all reported $Ru(bpy)_3^{2+}$-labeled ECL [14] with a linear range of 10^{-13} to 10^{-7} M in acetonitrile–H2O mixed solution containing persulfate. Composition of electrode has profound effect on ECL intensity, as reaction occurs in the proximity of electrode. Quenching by persulfate can be slow down by using carbon paste electrodes as they are made of carbon materials and organic solvents and can improve the stability of electrogenerated $Ru(bpy)_3^+$ by supplying organic phase at the electrode surface. To overcome the problem of serious hydrogen evolution upon applying very negative potentials, enhanced ECL emission in purely aqueous solution has been observed by using carbon paste electrode. Storage time of carbon paste electrode sometimes takes part in enhancement of ECL intensity. Amazingly, the ECL emission from $Ru(bpy)_3^{2+}$ and persulfate is greater at freshly prepared carbon paste electrode, whereas the ECL emission intensity from $Ru(phen)_3^{2+}$ and persulfate was not affected by the storage time [15]. Xu's group while studying ECL emission in aqueous solution by using bismuth electrode realized that the quenching effect of high concentration persulfate was weak at bismuth electrode. Owing to their high hydrogen overpotential, bismuth electrodes are expected to be powerful electrodes for cathodic ECL studies in aqueous solutions [16]. Moreover, the ECL emission of persulfate can also be observed in aqueous solutions by using easily reducible metal complexes $Ru(bpy)_3^{2+}$ or $Cr(bpy)_3^{3+}$ [17].

2.2.2 *"Oxidative–Reduction" Co-reactants*

2.2.2.1 Oxalate System

Bard and co-workers in 1977 observed ECL emission from oxalate in acetonitrile [8]. Later, they observed it in aqueous solutions following mechanism as shown in Eqs. (2.13)–(2.19) for the detection of oxalate and $Ru(bpy)_3^{2+}$ [10, 14]. In the following years, a fiber-optic-based ECL sensor for oxalate detection was constructed [18], and the reaction mechanism of the oxalate system was further investigated and applied to study the electron transfer at liquid/liquid interface [19]. Simultaneous electrochemical and ECL detection of oxalate was reported in 2000 by Forster et al. [20].

$$Ru(bpy)_3^{2+} - e^- \rightarrow Ru(bpy)_3^{3+} \qquad (2.13)$$

$$Ru(bpy)_3^{3+} + C_2O_4^{2-} \rightarrow Ru(bpy)_3^{2+} + C_2O_4^{\bullet-} \qquad (2.14)$$

$$C_2O_4^{\bullet-} \rightarrow CO_2^{\bullet-} + CO_2 \qquad (2.15)$$

$$Ru(bpy)_3^{3+} + CO_2^{\bullet-} \rightarrow Ru(bpy)_3^{2+*} + CO_2 \qquad (2.16)$$

or

$$Ru(bpy)_3^{2+} + CO_2^{\bullet-} \rightarrow Ru(bpy)_3^{+} + CO_2 \qquad (2.17)$$

$$Ru(bpy)_3^{+} + Ru(bpy)_3^{3+} \rightarrow Ru(bpy)_3^{2+} + Ru(bpy)_3^{2+*} \qquad (2.18)$$

$$Ru(bpy)_3^{2+*} \rightarrow Ru(bpy)_3^{2+} + h\nu(610 \text{ nm}) \qquad (2.19)$$

A clear ECL response was reported from the $[Ru(bpy)_2dcb]^{2+}$ SAMs deposited on optically transparent fluorine-doped tin oxide (FTO) electrodes in the presence of oxalate and amino acids as co-reactants (dcb = 4,4-dicarboxy-2,2-bipyridine) [21]. Few years back, Forster and co-workers studied the use of metallopolymers as a novel ECL platform using oxalate, TPA, and other small molecules as co-reactants [22] Ru(II) diimine complexes having phosphonic acid substituents adsorbed to TiO2-modified ITO electrodes can undergo co-reactant ECL with $C_2O_4^{2-}$ [23]. In order to improve the performance of existing systems and designing new ECL systems, a series of ruthenium bipyridine complexes and oxalate as a co-reactant were used to study the structure–activity relationships.

ECL generated by oxidation of $Ru(bpy)_3^{2+}$, $Ru(phen)_3^{2+}$, $Ru(bpy)_2(dmbp)^{2+}$, Ru-$(dmphen)_3^{2+}$, or $Ru(dmbp)_3^{2+}$ (dmbp) 4,4'-dimethyl-2,2'-bipyridine (dmphen), 4,7-dimethyl-1,10-phenanthroline, and oxalate in aqueous solution was attributed to the driving force for the electron-transfer reactions and the different pathways for co-reactant (i.e., $CO2^{\bullet-}$) reaction [19].

Some other papers reported microenvironmental effects of micelles on the electrochemical and ECL behavior of $Os(bpy)_3^{2+}$ [24]. For the one electron oxidation of $Os(bpy)_3^{2+/3+}$ near +0.6 V, aqueous solution using oxalate was used to generate $Os(bpy)_3^{2+*}$. A study involving electron/hole annihilation through electron transfer between sterically stabilized silicon nanocrystals (NCs) and redox-active co-reactants was reported. Higher ECL intensity light emission was observed from the NC solutions in the presence of oxalate or persulfate as co-reactant [25]. Oxalate and peroxydisulfate in aqueous solution were determined at subpicomolar levels [14] based on $Ru(bpy)_3^{2+}$ to generate ECL and are applied for the selective determination of oxalate synthetic urine samples [26]. The role of oxalate direct oxidation to the overall ECL behavior is dependent on many factors, e.g., surface property of the electrode, the concentration of oxalate, and the electrode potential applied [19, 27, 28]. ECL generation is also very sensitive to the solution pH. A maximum of pH $\sim$ 6 in unbuffered (aqueous) solutions at Pt [4], constant from pH 4 to 8 in phosphate buffer solutions at GC [14, 29], and from pH 5 to 8 in phosphate buffer solutions at ultramicroelectrodes [19] has been described for $Ru(bpy)_3^{2+}$/oxalate system.

2.2.2.2 Pyruvate System

Another example of the "oxidative–reduction" system is the determination of pyruvate by $Ru(bpy)_3^{2+}$, which was reported by Knight and Greenway. ECL from pyruvate either at the electrode or by electrogenerated $Ru(bpy)_3^{3+}$ cannot be observed, as the oxidation of pyruvate is suppressed by the presence of an oxide layer on the Pt electrode surface. At a reduced Pt electrode, the difference in pyruvate and $Ru(bpy)_3^{2+}$ oxidation potentials is negligibly small. As a result, the electrochemically produced $Ru(bpy)_3^{3+}$ is not strong enough to oxidize the pyruvate in solution, so no ECL is generated. Addition of Ce(III) to an acidic solution containing $Ru(bpy)_3^{2+}$ and pyruvate facilitates the ECL reaction upon anodic oxidation. Ce(IV) produced as a result of the oxidation of Ce(III) at the Pt electrode at higher potentials than $Ru(bpy)_3^{2+}$ oxidation is a strong oxidant and can effectively oxidize pyruvate as shown in Eqs. (2.20)–(2.25), ensuing in the formation of the strongly reducing intermediate $CH_3CO^{\bullet}$. This species behaves like $CO_2^{\bullet-}$ and participates in electron-transfer reactions with $Ru(bpy)_3^{3+}$ and $Ru(bpy)_3^{2+}$, resulting in the emission of ECL [30].

$$Ce^{3+} - e^- \rightarrow Ce^{4+} \tag{2.20}$$

$$Ce^{4+} + CH_3COCO_2^- \rightarrow Ce^{3+} + CH_3COCO_2^{\bullet} \tag{2.21}$$

$$CH_3COCO_2^{\bullet} \rightarrow CH_3CO^{\bullet} + CO_2 \tag{2.22}$$

$$Ru(bpy)_3^{2+} - e^- \rightarrow Ru(bpy)_3^{3+} \tag{2.23}$$

$$Ru(bpy)_3^{3+} + CH_3CO^{\bullet} \rightarrow Ru(bpy)_3^{2+*} + carbonium\ ion \tag{2.24}$$

$$Ru(bpy)_3^{2+*} \rightarrow Ru(bpy)_3^{2+} + hv(610\ nm) \tag{2.25}$$

2.2.2.3 Amine System

$Ru(bpy)_3^{2+}$/tripropylamine (TPA) system, the most common ECL system, is currently the basis for the commercial ECL immunoassays and DNA analyses. Its mechanism has been explained in detail in recent years. There are four distinct possible reaction mechanisms [31, 32] of this system as shown in Fig. 2.2. During a detailed study on reaction mechanisms, it was revealed that there are some difficulties with these mechanisms, such as the lifetime of the excited species (R_3N^+ cation) generated by oxidation seemed to be either too long or too short on the basis of electron paramagnetic resonance measurements and on the basis of voltammetric measurements [1]. Possibilities related to pH values were also discussed by Pastore [15]. As for pH values less than 5, the deprotonation of the R_3NH^+ was the rate-limiting step, whereas above pH 5, the formation of an amine neutral radical was the rate-limiting step [33]. In Scheme 1 of Fig. 2.2, the ECL reactions occur as follows (Eqs. 2.26–2.30). Oxidation of both $Ru(bpy)_3^{2+}$ and TPA takes place at the electrode surface, and $Ru(bpy)_3^{3+}$ is reduced by TPA to produce the excited state.

$$Ru(bpy)_3^{2+} - e^- \rightarrow Ru(bpy)_3^{3+} \tag{2.26}$$

$$TPA - e^- \rightarrow TPA^{\bullet+} \tag{2.27}$$

$$TPA^{\bullet+} \rightarrow TPA^{\bullet} + H^+ \tag{2.28}$$

$$Ru(bpy)_3^{3+} + TPA^{\bullet} \rightarrow Ru(bpy)_3^{2+*} + products \tag{2.29}$$

$$Ru(bpy)_3^{2+*} \rightarrow Ru(bpy)_3^{2+} + hv \tag{2.30}$$

In Scheme 2 of Fig. 2.2, $Ru(bpy)_3^{2+}$ is reduced by $TPA^{\bullet}$, and $Ru(bpy)_3^+$ reacts directly with $Ru(bpy)_3^{3+}$ to form the excited state, ensuing the emission of light. In Scheme 3 of Fig. 2.2, only $Ru(bpy)_3^{3+}$ is generated at the electrode surface, and TPA is oxidized by the generated $Ru(bpy)_3^{3+}$. $Ru(bpy)_3^{2+}$ concentration is a factor on which the overall ECL intensity depends; hence, this process is not favoured when $Ru(bpy)_3^{2+}$ concentrations are low. The process of Scheme 4 of Fig. 2.2 shows the direct oxidation of TPA at the electrode to generate $TPA^{\bullet+}$ and $TPA^{\bullet}$. The following reaction between $TPA^{\bullet}$ and $Ru(bpy)_3^{2+}$ generates $Ru(bpy)_3^+$, which in turn reacts with $TPA^{\bullet+}$ to form the excited-state $Ru(bpy)_3^{2+*}$. The ECL route in

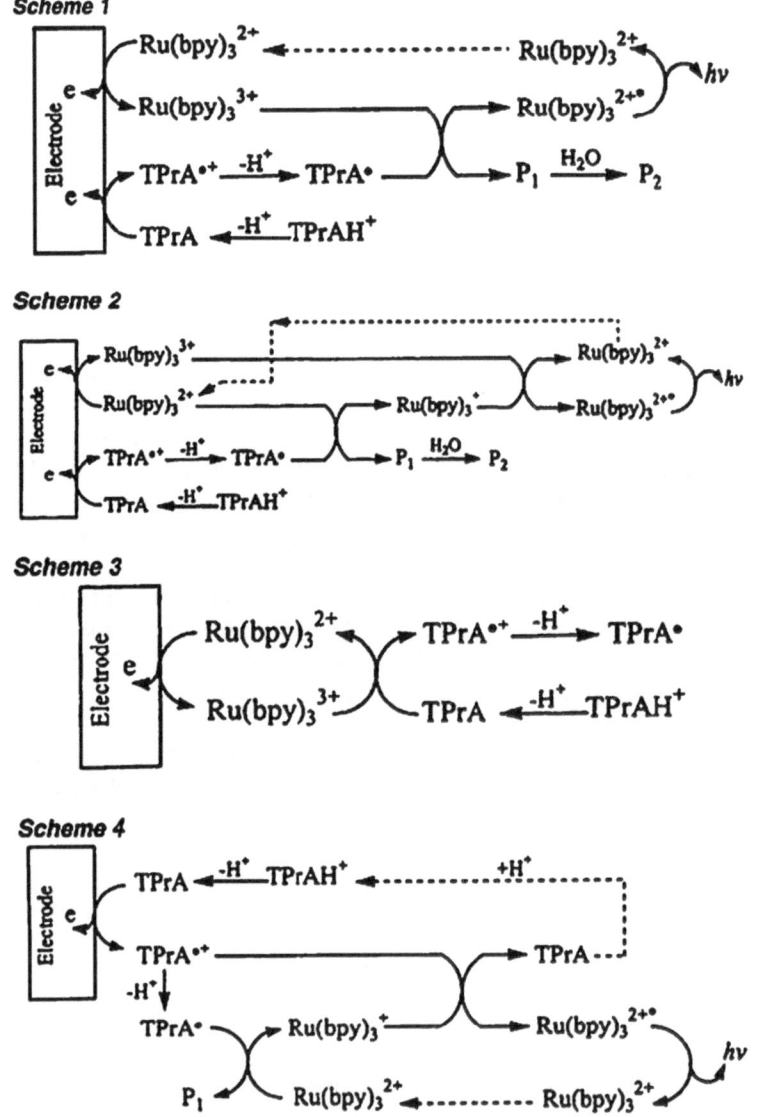

Fig. 2.2 Schemes for the reaction mechanisms of the Ru(bpy)$_3^{2+}$/TPA system (Reprinted with permission from Ref. [13]. Copyright 2002 American Chemical Society)

this Scheme can occur at an LOP that is just positive enough to oxidize TPA, but not enough to oxidize Ru(bpy)$_3^{2+}$, and is designated as LOP ECL [13]. Figure 2.3 illustrates tri-n-propylamine oxidation reaction sequence [34].

Further studies on the ECL intensity of the Ru(bpy)$_3^{2+}$/TPA system reveals that it strongly depends on the solution pH. A large number of ECL applications in biological systems appear in literature, owing to its feature that the maximum value of

Fig. 2.3 Proposed tri-*n*-propylamine oxidation/reaction sequence with abbreviations in parentheses (Reprinted with permission from Ref. [34]. Copyright 2004 American Chemical Society)

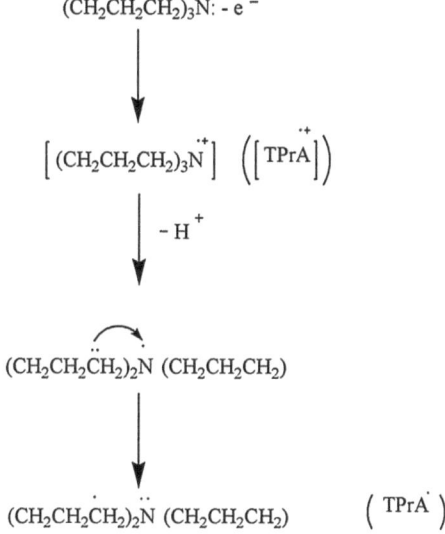

the ECL intensity occurs at around pH 7.5, which is exceptionally suitable for applications in biological systems. Sensitivity of the system may also be significantly increased by the addition of some surface-active agents [35–38]. These agents are in fact adsorbed on the surface of electrodes by which electrode surface becomes more hydrophobic, thus facilitating the direct oxidation of TPA. These results are supported by the effects of non-ionic chain length on $Ru(bpy)_3^{2+}$/TPA ECL [35] and the dependence of the surfactant effect on the electrode materials [36].

Non-ionic surfactants, fluorosurfactants [37], and the ionic surfactants [38] can also enhance ECL intensity up to 8-fold, 50-fold, and 30-fold, respectively. Hydroxylic solvents including fluorinated alcohols also play important role to enhance $Ru(bpy)_3^{2+}$/TPA ECL intensity remarkably [39]. Due to increase in polarity, hydrogen bonding and dipole forces dramatically change the ground- and excited-state properties in $Ru(bpy)_3^{2+}$, thus increasing the energy of maximum PL and ECL emission. Moreover, dramatic increases in ECL efficiencies using mixed alcohol/water solutions were observed ranging from 6- to 270-folds. Solution pH also affects the ECL intensity of $Ru(bpy)_3^{2+}$/TPA system [40, 41]. It shows dramatic increase in pH till pH > 5.5 with maximum pH value at 7.5. The exact reason of this phenomena might be indistinct; however, causes related to the deprotonation reactions of $TPAH^+$ and $TPA^{\bullet+}$, the stability of the intermediates formed, and the solubility decrease in TPA at high pH are also reported. Moreover, pH values higher than 9 should be avoided, as at this pH a significant ECL background signal is produced by $Ru(bpy)_3^{3+}$ reaction with hydroxide ions at electrode. ECL behavior of $Ru(bpy)_3^{2+}$/TPA in MeCN shows that the ECL generation in this case is predominately associated with the direct oxidation of $Ru(bpy)_3^{2+}$ at the electrode [42–45]. ECL with TPA in fluorinated and non-fluorinated alcohols and alcohol/water mixtures is also found in literature [39].

A number of articles [17, 34, 41, 46–49] have been published for the last few years for describing the correlation of ECL efficiency with the amine structure. Generally, ECL intensity enhancement occurs in the order primary < secondary < tertiary amines. For the amines to be ECL efficient, these should have α-hydrogen, so that upon oxidation, newly formed radical cation species can deprotonate easily to form a strongly reducing free radical species [41]. The nature of substituents and other functional groups on amine molecule also plays role in determining the ECL intensity. Usually, electron-withdrawing substituents tend to cause a reduction in ECL activity, and electron-donating groups have the opposite effect. Aromatic amines, aromatic substituted amines, and amines with a carbon–carbon double bond that can conjugate the radical intermediates consistently give a very low ECL response [50]. The ECL reaction mechanism of $Ru(bpy)_3^{2+}$ with six tertiary aliphatic amines, namely tri-n-butylamine (TBuA), tri-isobutylamine (TisoBuA), TPA, methyl-di-n-propylamine (MeDPrA), triethylamine (TEtA), and trimethylamine (TMeA), was studied in aqueous solution, and their ECL was examined using fast potential pulses at carbon-fiber microelectrodes and with simulation techniques to obtain information on the E° value of the amine redox couples [33].

Although having marvelous features and attractiveness, TPA has several short-comings such as toxicity and volatility, and high concentrations (usually up to 100 mM) are needed to be used to obtain good sensitivity. 2-(dibutylamino)ethanol (DBAE), a more environmentally-friendly co-reactant was introduced with approximately 10 and 100 times greater ECL enhancement at Au and Pt electrodes, respectively [51] (Fig. 2.4). This improved ECL intensity was attributed to the catalytic effect of hydroxyl group toward the direct oxidation of DBAE at the electrode. The sensitivity of the $Ru(bpy)_3^{2+}$/DBAE system is about an order of magnitude better than that of the $Ru(bpy)_3^{2+}$/TPA system. DBAE is much less toxic and less volatile than TPA and more soluble in aqueous solutions, and 20 mM DBAE is more effective than 100 mM TPA. These features make DBAE a very promising co-reactant for $Ru(bpy)_3^{2+}$ ECL immunoassay and DNA probe assays. N-butyldi-ethanolamine containing two hydroxyl groups was found even more effective than DBAE containing one hydroxyl at Au and Pt electrodes and is the most effective co-reactant reported so far. While in search of more efficient co-reactants, it was revealed that monoamines exhibited much higher ECL than the diamines, which may result from considerable side reactions of diamines [52]. These results may be helpful for new chemists in investigating new highly efficient co-reactants.

2.3 Hot-Electron ECL or Cathodic Luminescence

ECL can also be generated via hot electrons emitted into electrolyte solution from an oxide-covered electrode surface, sometimes called a conductor/insulator/elec-trolyte surface [53]. This special light emission pathway in ECL is commonly known as hot-electron-induced ECL [54–56], where hot electrons are defined as

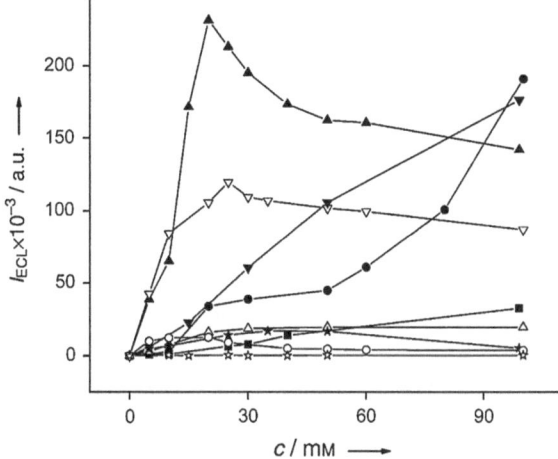

Fig. 2.4 Dependence of the ECL peak intensity on the concentrations of 2-(dibutylamino)eth-anol (DBAE, *filled triangle*), triethanolamine (*empty square*), N,N-diethylethanolamine (*filled inverted triangle*), TPA (*filled circle*), triethylamine (*filled square*), N,N-diethyl-N'-methylethy-lenediamine (*empty triangle*), 3-diethylamino-1-propanol (*filled stars*), nitrilotriacetic acid (*empty circle*), and ethylenediaminetetraacetic acid (*empty stars*), measured at the Au electrode in 0.1 m phosphate buffer solution (pH 7.5) containing 1 mm [Ru(bpy)$_3$]$^{2+}$. The potential was stepped from 0 to 1.35 V (Reprinted with permission from Ref. [51]. Copyright 2007 Wiley-VCH)

electrons possessing thermal energies greater than the thermal energy of a phase or as electrons at energy far above the Fermi energy of a phase [57, 58].

The insulating layer of metal oxide used for generation of hot electrons prevents electron transferring from the metal to a solution species, until a large electric field is present across the oxide and the Fermi level in the metal is above that of the conduction band of the metal oxide. Current flow across the oxide film can then produce a hot solution-phase electron [56]. Strong reductants are generated due to the strong reductive ability of the hot electrons which then react with the emitter to generate an excited state and then emit light. For example, both Ru(bpy)$_3$$^{2+}$ and luminol exhibit strong hot-electron-induced ECL on an oxide-covered aluminum electrode based on a tunnel emission of hot electrons into an aqueous solution [56, 59]. By using this technique, subnanomolar detection of Ru(bpy)$_3$$^{2+}$ and luminol labels with a dynamic range of several orders of magnitude is realized. ECL emission of Ru(bpy)$_3$$^{2+}$ or the mixed solution of Ru(bpy)$_3$$^{2+}$ and persulfate on oxide-covered aluminum electrode was described by Kulmala et al. [52].

Soon afterward, a detection method for Ru(bpy)$_3$$^{2+}$ based on the ECL of Ru(bpy)$_3$$^{2+}$ and persulfate on oxide-covered aluminum electrode [60] and Ru(bpy)$_3$$^{3+}$ in acetonitrile and aqueous solution on oxide-covered tantalum electrode was developed [11]. The ECL intensity showed linearity in the concentration range of 10^{-10} M and 10^{-5} M Ru(bpy)$_3$$^{2+}$ in the presence of 1 mM persulfate. The sensitivity was improved by an order of magnitude in the presence of Tween 20, and this technique was applied to immunoassays. The ECL emission on oxide-covered aluminum or tantalum electrode may be generated by the reaction between $SO_4^{\bullet-}$ and $Ru(bpy)_3^+$, the reaction between hot electron and $Ru(bpy)_3^{3+}$, and the reaction between $Ru(bpy)_3^+$ and $Ru(bpy)_3^{3+}$ (Eqs. 2.39–2.45) [60].

$$e_{aq}^- + S_2O_8^{2-} \rightarrow SO_4^{\bullet-} + SO_4^{2-} \tag{2.39}$$

$$Ru(bpy)_3^{2+} + e_{aq}^- \rightarrow Ru(bpy)_3^+ \tag{2.40}$$

$$Ru(bpy)_3^{2+} + SO_4^{\bullet-} \rightarrow Ru(bpy)_3^{3+} + SO_4^{2-} \tag{2.41}$$

$$Ru(bpy)_3^+ + SO_4^{\bullet-} \rightarrow Ru(bpy)_3^{2+*} + SO_4^{2-} \tag{2.42}$$

$$Ru(bpy)_3^{3+} + e_{aq}^- \rightarrow Ru(bpy)_3^{2+*} \tag{2.43}$$

$$Ru(bpy)_3^{3+} + Ru(bpy)_3^+ \rightarrow Ru(bpy)_3^{2+*} + Ru(bpy)_3^{2+} \tag{2.44}$$

$$Ru(bpy)_3^{2+*} \rightarrow Ru(bpy)_3^{2+} + h\nu \tag{2.45}$$

References

1. Forster RJ, Bertoncello P, Keyes TE (2009) Electrogenerated Chemiluminescence. Ann Rev Anal Chem 2:359–385. doi:10.1146/annurev-anchem-060908-155305
2. Miao W (2008) Electrogenerated chemiluminescence and its biorelated applications. Chem Rev 108(7):2506–2553. doi:10.1021/cr068083a
3. Bard AJ, Faulkner LR (2001) Electrochemical methods: fundamentals and applications, 2nd edn. Wiley, New York
4. Rubinstein I, Bard AJ (1981) Electrogenerated chemiluminescence. 37. Aqueous ECL systems based on tris(2,2'-bipyridine)ruthenium($2+$) and oxalate or organic acids. J Am Chem Soc 103(3):512–516. doi:10.1021/ja00393a006
5. Bertoncello P, Forster RJ (2009) Nanostructured materials for electrochemiluminescence (ECL)-based detection methods: Recent advances and future perspectives. Biosens Bioelectron 24(11):3191–3200. doi:10.1016/j.bios.2009.02.013
6. Knight AW, Greenway GM (1994) Occurrence, mechanisms and analytical applications of electrogenerated chemiluminescence: review. Analyst 119(5):879–890. doi:10.1039/an9941900879
7. Tokel NE, Bard AJ (1972) Electrogenerated chemiluminescence. IX. Electrochemistry and emission from systems containing tris(2,2'-bipyridine)ruthenium(II) dichloride. J Am Chem Soc 94(8):2862–2863. doi:10.1021/ja00763a056
8. Chang M-M, Saji T, Bard AJ (1977) Electrogenerated chemiluminescence. 30. Electrochemical oxidation of oxalate ion in the presence of luminescers in acetonitrile solutions. J Am Chem Soc 99(16):5399–5403. doi:10.1021/ja00458a028
9. Michel PE, de Rooij NF, Koudelka-Hep M, Fähnrich KA, O'Sullivan CK, Guilbault GG (1999) Redox-cycling type electrochemiluminescence in aqueous medium. A new principle for the detection of proteins labeled with a ruthenium chelate. J Electroanal Chem 474(2):192–194. doi:10.1016/S0022-0728(99)00351-4
10. Li H-J, Han S, Hu L-Z, Xu G-B (2009) Progress in Ru (bpy)($_3$)($2+$) Electrogenerated Chemiluminescence. Chin J Anal Chem 37(11):1557–1565

11. White HS, Bard AJ (1982) Electrogenerated chemiluminescence. 41. Electrogenerated chemiluminescence and chemiluminescence of the $Ru(2,21\text{-bpy})_3^{2+}\text{-}S_2O_8^{2-}$ system in acetonitrile-water solutions. J Am Chem Soc 104(25):6891–6895. doi:10.1021/ja00389a001

12. Miao W, Choi JP (2004) Electrogenerated Chemiluminescence. In: Marcel Dekker, New York, p 213

13. Hu L, Xu G (2010) Applications and trends in electrochemiluminescence. Chem Soc Rev 39(8):3275–3304. doi:10.1039/b923679c

14. Ege D, Becker WG, Bard AJ (1984) Electrogenerated chemiluminescent determination of tris(2,2′-bipyridine) ruthenium ion $\left(Ru(bpy)^3{}_{2+}\right)$ at low levels. Anal Chem 56(13):2413–2417. doi:10.1021/ac00277a036

15. Gorman BA, Francis PS, Barnett NW (2006) Tris(2,2′-bipyridyl)ruthenium(II) chemiluminescence. Analyst 131(5):616–639. doi:10.1039/b518454a

16. Hu L, Li H, Zhu S, Fan L, Shi L, Liu X, Xu G (2007) Cathodic electrochemiluminescence in aqueous solutions at bismuth electrodes. Chem Comm 0(40):4146–4148

17. Fahnrich KA, Pravda M, Guilbault GG (2001) Recent applications of electrogenerated chemiluminescence in chemical analysis. Talanta 54(4):531–559. doi:10.1016/s0039-9140(01)00312-5

18. Egashira N, Kumasako H, Ohga K (1990) Fabrication of a fiber optic based electrochemiluminescence sensor and its application to the determination of oxalate. Anal Sci 6:903–904

19. Kanoufi F, Bard AJ (1999) Electrogenerated Chemiluminescence. 65. An investigation of the oxidation of oxalate by tris(polypyridine) ruthenium complexes and the effect of the electrochemical steps on the emission intensity. J Phy Chem B 103(47):10469–10480. doi:10.1021/jp992368s

20. Forster RJ, Hogan CF (2000) Electrochemiluminescent metallopolymer coatings: Combined light and current detection in flow injection analysis. Anal Chem 72(22):5576–5582. doi:10.1021/ac000605d

21. Dennany L, O'Reilly EJ, Keyes TE, Forster RJ (2006) Electrochemiluminescent monolayers on metal oxide electrodes: Detection of amino acids. Electrochem Comm 8(10):1588–1594. doi:10.1016/j.elecom.2006.07.022

22. Dennany L, Hogan CF, Keyes TE, Forster RJ (2006) Effect of surface immobilization on the electrochemiluminescence of ruthenium-containing metallopolymers. Anal Chem 78(5):1412–1417. doi:10.1021/ac0513919

23. Andersson A-M, Isovitsch R, Miranda D, Wadhwa S, Schmehl RH (2000) Electrogenerated chemiluminescence from Ru() bipyridylphosphonic acid complexes adsorbed to mesoporous TiO/ITO electrodes. Chem Comm 0(6):505–506

24. Bard AJ (2004) Electrogenerated Chemiluminescence. Taylor & Francis

25. Ding Z, Quinn BM, Haram SK, Pell LE, Korgel BA, Bard AJ (2002) Electrochemistry and electrogenerated chemiluminescence from silicon nanocrystal quantum dots. Science (New York, NY) 296(5571):1293–1297. doi:10.1126/science.1069336

26. Rubinstein I, Martin CR, Bard AJ (1983) Electrogenerated chemiluminescent determination of oxalate. Anal Chem 55(9):1580–1582. doi:10.1021/ac00260a030

27. Li F, Cui H, Lin X-Q (2002) Potential-resolved electrochemiluminescence of $Ru(bpy)_3{}^{2+}$/$C2O42^-$ system on gold electrode. Luminescence 17(2):117–122. doi:10.1002/bio.674

28. Lu M-C, Whang C-W (2004) The role of direct oxalate oxidation in electrogenerated chemiluminescence of poly(4-vinylpyridine)-bound $Ru(bpy)_2Cl^+$/oxalate system on indium tin oxide electrodes. Anal Chim Acta 522(1):25–33. doi:10.1016/j.aca.2004.06.042

29. Downey TM, Nieman TA (1992) Chemiluminescence detection using regenerable tris(2,2′-bipyridyl)ruthenium(II) immobilized in Nafion. Anal Chem 64(3):261–268. doi:10.1021/ac00027a005

30. Knight AW, Greenway GM (1995) Indirect, ion-annihilation electrogenerated chemiluminescence and its application to the determination of aromatic tertiary-amines. Analyst 120(4):1077–1082. doi:10.1039/an9952001077

31. Miao W, Choi J-P, Bard AJ (2002) Electrogenerated Chemiluminescence 69: The Tris(2,2'-bipyridine)ruthenium(II), (Ru(bpy)$_3$$^{2+}$)/Tri-n-propylamine (TPrA) system revisited. A new route involving tpra•+ cation radicals. J Am Chem Soc 124(48):14478–14485. doi:10.1021/ja027532v

32. Wightman RM, Forry SP, Maus R, Badocco D, Pastore P (2004) Rate-Determining Step in the Electrogenerated Chemiluminescence from Tertiary Amines with Tris(2,2'-bipyridyl)ruthenium(II). J Phy Chem B 108(50):19119–19125. doi:10.1021/jp036034l

33. Pastore P, Badocco D, Zanon F (2006) Influence of nature, concentration and pH of buffer acid–base system on rate determining step of the electrochemiluminescence of Ru(bpy)$_3$2$^+$ with tertiary aliphatic amines. Electrochim Acta 51(25):5394–5401. doi:10.1016/j.electacta.2006.02.009

34. Richter MM (2004) Electrochemiluminescence (ECL). Chem Rev 104(6):3003–3036. doi:10.1021/cr020373d

35. Factor B, Muegge B, Workman S, Bolton E, Bos J, Richter MM (2001) Surfactant chain length effects on the light emission of tris(2,2'-bipyridyl)ruthenium(II)/tripropylamine electrogenerated chemiluminescence. Anal Chem 73(19):4621–4624

36. Zu YB, Bard AJ (2001) Electrogenerated chemiluminescence. 67. Dependence of light emission of the tris(2,2')bipyridylruthenium(II)/tripropylamine system on electrode surface hydrophobicity. Anal Chem 73(16):3960–3964. doi:10.1021/ac010230b

37. Li F, Zu Y (2004) Effect of nonionic fluorosurfactant on the electrogenerated chemiluminescence of the tris(2,2'-bipyridine)ruthenium(II)/tri-n-propylamine system: lower oxidation potential and higher emission intensity. Anal Chem 76(6):1768–1772. doi:10.1021/ac035181c

38. Xu G, Pang H-L, Xu B, Dong S, Wong K-Y (2005) Enhancing the electrochemiluminescence of tris(2,2[prime or minute]-bipyridyl)ruthenium(ii) by ionic surfactants. Analyst 130(4):541–544

39. Vinyard DJ, Richter MM (2007) Enhanced Electrogenerated Chemiluminescence in the Presence of Fluorinated Alcohols. Anal Chem 79(16):6404–6409. doi:10.1021/ac071028x

40. Leland JK, Powell MJ (1990) Electrogenerated chemiluminescence: an oxidative-reduction type ECL reaction sequence using Tripropyl Amine. J Electrochem Soc 137(10):3127–3131. doi:10.1149/1.2086171

41. Knight AW, Greenway GM (1996) Relationship between structural attributes and observed electrogenerated chemiluminescence (ECL) activity of tertiary amines as potential analytes for the tris(2,2-bipyridine)ruthenium(II) ECL reaction. A review. Analyst 121(11):101R–106R

42. Miao W, Bard AJ (2004) Electrogenerated chemiluminescence. 80. C-reactive protein determination at high amplification with [Ru(bpy)$_3$]$^{2+}$-containing microspheres. Anal Chem 76(23):7109–7113. doi:10.1021/ac048782s

43. Miao W, Bard AJ (2004) Electrogenerated chemiluminescence. 77. DNA hybridization detection at high amplification with [Ru(bpy)$_3$]$^{2+}$-containing microspheres. Anal Chem 76(18):5379–5386. doi:10.1021/ac0495236

44. Fan F-RF, Cliffel D, Bard AJ (1998) Scanning Electrochemical Microscopy. 37. Light emission by electrogenerated chemiluminescence at SECM tips and their application to scanning optical microscopy. Anal Chem 70(14):2941–2948. doi:10.1021/ac980107t

45. Richter MM, Bard AJ, Kim W, Schmehl RH (1998) Electrogenerated chemiluminescence. 62. Enhanced ECL in bimetallic assemblies with ligands that bridge isolated chromophores. Anal Chem 70(2):310–318. doi:10.1021/ac970736n

46. Andersson AM, Schmehl RH (2001) Molecular and supramolecular photochemistry 7: Optical sens switches

47. Gerardi RD, Barnett NW, Lewis SW (1999) Analytical applications of tris(2,2'-bipyridyl) ruthenium(III) as a chemiluminescent reagent. Anal Chim Acta 378(1–3):1–41. doi:10.1016/s0003-2670(98)00545-5

48. Knight AW (1999) A review of recent trends in analytical applications of electrogenerated chemiluminescence. Trac-Trends Anal Chem 18(1):47–62. doi:10.1016/s0165-9936(98)00086-7

49. Lee WY (1997) Tris (2,2'-bipyridyl)ruthenium(II) electrogenerated chemiluminescence in analytical science. Mikrochim Acta 127(1–2):19–39. doi:10.1007/bf01243160

50. Bock CR, Connor JA, Gutierrez AR, Meyer TJ, Whitten DG, Sullivan BP, Nagle JK (1979) Estimation of excited-state redox potentials by electron-transfer quenching. Application of electron-transfer theory to excited-state redox processes. J Am Chem Soc 101(17):4815–4824. doi:10.1021/ja00511a007

51. Liu X, Shi L, Niu W, Li H, Xu G (2007) Environmentally friendly and highly sensitive ruthenium(ii) tris(2,2'-bipyridyl) electrochemiluminescent system using 2-(dibutylamino)-ethanol as co-reactant. Angew Chem Int Ed 46(3):421–424. doi:10.1002/anie.200603491

52. Han S, Niu W, Li H, Hu L, Yuan Y, Xu G (2010) Effect of hydroxyl and amino groups on electrochemiluminescence activity of tertiary amines at low tris(2,2'-bipyridyl)ruthenium(II) concentrations. Talanta 81(1–2):44–47. doi:10.1016/j.talanta.2009.11.037

53. Pyati R, Richter MM (2007) ECL-Electrochemical luminescence. Ann Rep Sec "C" (Phy Chem) 103(0):12–78

54. Kankare J, Fäldén K, Kulmala S, Haapakka K (1992) Cathodically induced time-resolved lanthanide(III) electroluminescence at stationary aluminium disc electrodes. Anal Chim Acta 256(1):17–28. doi:10.1016/0003-2670(92)85320-6

55. Kankare J, Haapakka K, Kulmala S, Näntö V, Eskola J, Takalo H (1992) Immunoassay by time-resolved electrogenerated luminescence. Anal Chim Acta 266(2):205–212. doi:10.1016/0003-2670(92)85044-7

56. Gaillard F, Sung Y-E, Bard AJ (1999) Hot electron generation in aqueous solution at oxide-covered tantalum electrodes. Reduction of methylpyridinium and electrogenerated chemiluminescence of $Ru(bpy)_3^{2+}$. J Phy Chem B 103(4):667–674. doi:10.1021/jp982821k

57. Nozik AJ (2001) Spectroscopy and hot electron relaxation dynamics in semiconductor quantum wells and quantum dots. Ann Rev Phy Chem 52(1):193–231. doi:10.1146/annurev.physchem.52.1.193

58. Verlet JRR (2008) Femtosecond spectroscopy of cluster anions: insights into condensed-phase phenomena from the gas-phase. Chem Soc Rev 37(3):505–517

59. Kulmala S, Ala-Kleme T, Kulmala A, Papkovsky D, Loikas K (1998) Cathodic electrogenerated chemiluminescence of luminol at disposable oxide-covered aluminum electrodes. Anal Chem 70(6):1112–1118. doi:10.1021/ac970954g

60. Xu G, Dong S (2000) Electrochemiluminescence of the $Ru(bpy)_3^{2+}/S2O82^-$ system in purely aqueous solution at carbon paste electrode. Electroanal 12(8):583–587. doi:10.1002/(SICI)1521-4109(200005)12

Chapter 3
ECL Instrumentation

Abstract The basic components of an ECL instrument consist of supply of an electrical energy for the ECL reaction at an electrode within an electrochemical cell and an optical detector for the measurement of the emitted light intensity. Although certain types of ECL instruments are now commercially available, most of the ECL studies were carried out in homemade ECL instruments according to most of the literature. A review of some new developments in the ECL instrumentation and ECL cells is described.

Keywords Nonaqueous electrochemical media · Electrochemiluminescent cell · Lab on a chip · Light detection · Charged coupled device cameras

As advantages are inherent in ECL technique, it is possible to remark the basic instrumentation required and to simplify the instrument needed. Including an electrochemical apparatus equipped with potentiostat (enabling two- or three-electrode system) is the basis of ECL instrumentation. The electrodes used are mainly platinum-based, while gold, carbon, and mercury are also used as electrode materials. Moreover, for optically transparent electrodes, a fine metal gauze or indium tin oxide (ITO) on quartz may be used. For sensitive ECL detection, one of the most important parameters is the shape of the electrochemical/ECL cell and its position in an absolutely dark box aligned according to a very sensitive photomultiplier or a camera. Furthermore, voltammetric curve (i versus E) and the dependence of the ECL intensity on potential (I_{ECL} versus E) should be simultaneously recorded. In the advent times of ECL, Fritz Pragst designed a simple instrument for the possibility to follow and to photograph the intensity of the emitted light and to record the respective spectra [1].

The basic components of an ECL instrument consist of (1) an electrical energy supply for the ECL reaction at an electrode and (2) an optical detector for the measurement of either the emitted light intensity (as photocurrent or counts s^{-1}) or its spectroscopic response [2, 3]. Although certain types of ECL instruments are now commercially available (as discussed in the next section), most of the ECL studies reported in the literature were performed in homemade ECL instruments.

3.1 Nonaqueous Electrochemical Media

Modern ECL instruments used for the detection of biorelated species are mostly carried out in aqueous media, while early reports display that historically all ECL studies except for luminol systems were conducted in organic media. In case of biorelated species, the purity of the solvent/supporting electrolyte plays a significant role. Trace amounts of water and oxygen present as an impurity either can disable the generation of both reductive and oxidative ECL precursor species at the electrode or may quench the newly formed excited state species. To avoid such conditions, the ECL experiment is often performed in an oxygen-free dry box with pure and dry solvent and electrolyte. Readers may find more details on solvent and electrolyte drying and purification [4, 5] and commonly used solvent-supporting electrolyte systems for ECL study [2, 3] in the literature.

3.2 Cell Design

ECL cell design depends upon the different types of ECL studies. Ion annihilation ECL studies require an oxygen- and moisture-free sealed system [2]. Suitably sized glass vials may be used as a cell for general co-reactant ECL studies, as degassing in this case is insignificant. While carrying out ECL experiments with a wafer-type electrode, the effective area of the electrode exposed to the co-reactant solution needs to be carefully controlled and be aligned to face the window of the photodetector.

Overwhelming majority of publications [6–9] concerning ECL reveals that it has been widely used as a detector in FIA, LC, CE, and microchip CE. For ECL equipped with FIA, various types of thin-layer flow-through ECL cells have been constructed; some of them are exemplified in Fig. 3.1 [10].

Several drawbacks are attached with these kinds of ECL cells including (1) large dead volumes which greatly affect detection sensitivity and separation efficiency of a sample, (2) high IR drops which decreases the sensitivity of ECL detection, and (3) high flow resistance which may cause significant noise due to difficulties of removing gas bubbles. To prevail over these problems, a new type of ECL flow cell is introduced comprising of a capillary inlet, a Pt ring working electrode, and a stainless steel pipe counter electrode. To prevent any undesired ECL emission from the electrode, the counter electrode is coated with a black plastic jacket and is used for the detection of 2-thiouracil in biological samples (Fig. 3.2) [11, 12].

Development of miniaturized analytical systems, the so-called lab on a chip, typically consist of a multi-layer glass, silicon, or polymer cell with small channels for the passage of reagents moved by conventional FI methods "off-chip", or by electro-osmosis "on-chip"[13]. Combined with such systems, ECL has great prospective as a detection technique. The electrodes are substituted as miniaturized

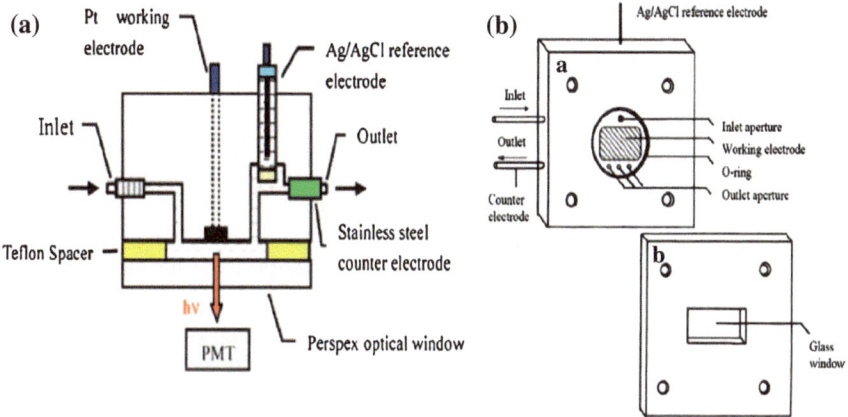

Fig. 3.1 Schematic diagrams of FI-ECL thin-layer flow-through cells **a** is the cell body and **b** the cell cover window (Reprinted with permission from Ref. [10]. Copyright 2008 American Chemical Society)

Fig. 3.2 Distribution of the thin layer at the tip of the Pt ring-immobilized capillary (Reprinted with permission from Ref. [12]. Copyright 2006 American Chemical Society)

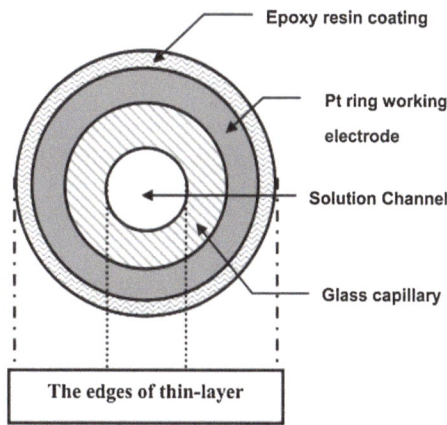

thin metal plates, fine wires, or metal coatings of inert substrates [14]. A platinum foil was sealed between two microscope slides with epoxy into an ECL flow cell to construct a microelectrode (w = 2 μm, l = 2.5 cm) which is used in an ECL cell for high-frequency ECL studies, providing a means for investigating ECL without rigorously purifying solvents or working on a vacuum line or in a dry box [15].

Printed circuit board (PCB) technology was employed for the fabrication of electrodes to be employed for luminol–H_2O_2 solutions. These electrodes were used with two microfluidic ECL cells that combine transparent polydimethylsiloxane (PDMS) microchannels [16]. The differences between the two cells were the working electrode size (10 mm^2 Au vs 0.09 mm^2 Au) and the ECL detection volume (4 μL vs 4 nL). No counter electrode was used in this system while "Ag/AgCl" reference electrode was obtained by electrodeposition of Ag on Au, followed by electro-oxidation of Ag in the presence of 0.10 M KCl. Noteworthily, it has also been reported recently that ECL using $Ru(bpy)_3^{2+}$ as luminophore and amino acids as co-reactants was determined employing a newly designed micro-fluidic cell to transport and mix two different solutions on the chip (Fig. 3.3). Such a cell requires only 50 nL of analyte solutions for ECL test. Drawback of this type of cell, however, is that it cannot distinguish one analyte from another if the sample solution contains several ECL-active unknown species [17].

For the design of the ECL cell for CE-ECL separation, a number of factors should be kept in mind [7, 8, 18]. LOD or elimination of the electrophoretic current from the ECL current is one of the foremost important factors. ECL reactions at the electrode are affected by the electrical current in capillaries under the high electric field. Increase in background signals happens due to change in the redox potential of the emitting species [e.g., $Ru(bpy)_3^{2+}$] and the interference of the electrode potential control of an amperometric controller [19, 20]. This elec-trophoretic current was minimized by using an on-column fracture covered with a Nafion tube [21], a porous polymer junction near the end of the capillary [22], a section of outside wall-etched capillary [23], and a porous etched joint capillary [18]. Sensitivity, the alignment or arrangement of the working electrode against

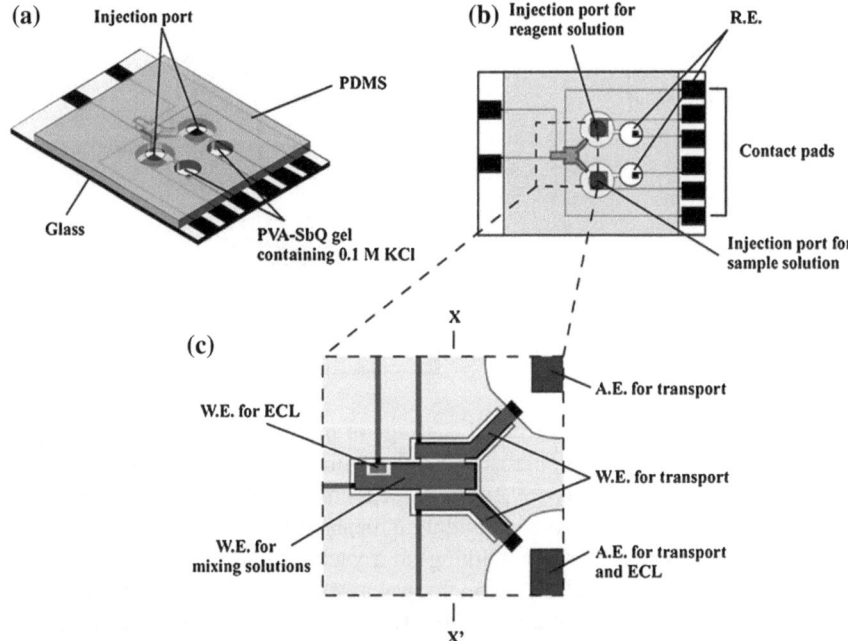

Fig. 3.3 Active microfluidic transport system with integrated components for optical sensing based on ECL: **a** completed chip; **b** planar layout of the entire system; **c** magnified view of the mixing area showing electrodes for the transport, mixing, and generation of the ECL. The dimensions of the chip are 14×20 mm. W.E., working electrode; R.E., reference electrode; A.E., auxiliary electrode (Reprinted with permission from Ref. [17]. Copyright 2006 Elsevier B.V)

the detection capillary and the photo detectors is also an important factor to be considered [24, 25]. Ideally, all of the separated analyte should reach the electrode and take part in the ECL reactions, for this, the distance between the end of the capillary and the working electrode needs to be optimized. In case of a shorter distance, $Ru(bpy)_3^{2+}$ diffusion is blocked, while too long a distance would dilute the analyte concentration at the electrode, resulting in poor sensitivity and reproducibility. Another factor is selectivity, the CE separation efficiency, influenced by the length of the detection capillary [18]. Because there is no electric field gradient in the detection capillary, the movement of the sample zone is governed by the pressure generated in the capillary. Thus, band broadening can occur by using a nonreasonable-sized detection capillary.

An ECL detection cell used for CE separation was described in a recent publication (Fig. 3.4), in which the joint was fabricated by etching the capillary wall with hydrofluoric acid after removal of half of the circumference of the polyimide coating in a 2–3-cm section [18]. The present ECL cell offers some advantages over previously reported ECL cells, such as the joint is quite strong and whole system is fabricated with no need to fix the joint on a plate. Band broadening effect is decreased by using a very short detection capillary of ~ 7 mm, hence increasing the CE efficiency. Moreover, the sample loss is small, and the part replacement is relatively easy [23].

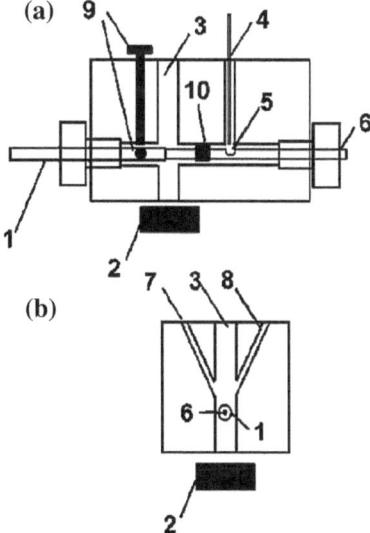

Fig. 3.4 Schematic diagram of the ECL detection cell coupled with capillary electrophoresis (CE) separation (not to scale): (**a**) front view and (**b**) side view from $Ru(bpy)_3^{2+}$ solution reservoir—(*1*) working electrode; (*2*) photomultiplier tube (PMT); (*3*) $Ru(bpy)_3^{2+}$ solution reservoir; (*4*) CE ground electrode; (*5*) porous section of capillary; (*6*) separation capillary; (*7*) reference electrode; (*8*) counter electrode; (*9*) working electrode alignment screws; (*10*) seal-on film [to isolate the $Ru(bpy)_3^{2+}$ solution reservoir and CE ground cell and to align the separation capillary] (Reprinted with permission from Ref. [18]. Copyright 2004 American Chemical Society)

Microchip CE-based separation with ECL detection is gaining more and more interest due to a number of advantages over classical CE, which include high performance, short analysis time, portability, disposability, and consumption of minute sample and reagent [7, 8, 26, 27]. An ECL probe, without the need of a conventional dark box to exclude ambient light [28], was used to take measurements in a similar fashion to a pH probe. Later on, another research group reported one-[29], two-[30], and three-channel microfluidic sensors [31] that were capable of detecting redox reactions indirectly using anodic co-reactant ECL [i.e., $Ru(bpy)_3^{2+}$/TPA system]. Activity and performance of all three proposed sensors are influenced by charge balance between the anode and cathode. In other words, the current at the cathode must equal the current at the anode. A simple and universal wall-jet configuration was constructed for the microchip CE-ECL detection system, to which both pre-column and post-column detection modes were applied to determine TPA and pharmaceuticals [32]. Contrary to the above CE-ECL devices, in which solution-phase $Ru(bpy)_3^{2+}$ was either added to the running buffer or placed in the detection reservoir, other approaches have also been employed, such as solid-state ECL detectors coupled with microchip CE by immobilizing $Ru(bpy)_3^{2+}$ into either Eastman AQ55D–silica–carbon nanotube composite thin film on a patterned ITO electrode [26] or zirconia–Nafion composite on a GC disk electrode [33] used for the detection of several pharmaceuticals.

3.3 Light Detection

In ECL, following three photo detectors are most frequently used, which include avalanche photodiodes (APD), photomultiplier tubes (PMTs), and charged coupled device (CCD) cameras. Among all these three, PMTs play exceptionally imperative role in sensitive detection of light and single photons [34]. For ECL studies, most commonly used PMTs include Hamamatsu [10] R4240 and R928 PMTs with spectral response ranges of 185–710 and 185–900 nm, respectively. These offer extremely high cathode sensitivity, low dark current and good signal-to-noise ratio. As R928 PMT has greater spectral response range (185–900 nm), it is thought that R928 PMT might give better results in case of $Ru(bpy)_3^{2+}$ species as compared to R4240. A new kind of PMT, Hamamatsu H742s1-40 metal package photon-counting PMT has been recently reported in the application of ECL study [35].

This PMT has some distinct features such as the PMT uses a GaAsP photocathode which displays a remarkably high quantum efficiency of 40 % at 580 nm and very low dark count of ~ 100–300 s^{-1}. CCD camera offers the advantages of instant image manipulation, high spatial resolution, and multi-channel detection ability. It has received increasing attention in ECL imaging and high throughput analysis [36]. To reduce dark current for enormously weak ECL measurements, such as in the ECL spectral recording experiments, the photodetector, namely PMT or CCD camera, often needs to be operated under low temperatures. Readers may take benefit of detailed discussions of this aspect in Refs. [2, 10, 37].

3.4 Commercial ECL Instruments

Although considerable interest is currently being focused on the development of certain types of ECL instruments, only a few of them are now commercially available. The fundamental experimental instrumentations of ECL study usually consist of two parts; the electrochemical device and the optical device. Light is detected by a PMT biased at a high voltage with a high-voltage power supply, a charged coupled device (CCD) camera, or a photodiode. In 1994, IGEN International, Inc. (technology later acquired by Roche Diagnostics Corp.) gained the title of introducing the first commercial ECL instrument (Fig. 3.5). Basically, this Origen analyzer is a flow cell-based system [10] where magnetic microbeads modified with probe molecules (antibodies or nucleic acids) were used to capture analyte molecules in a sample and $Ru(bpy)_3^{2+}$ labels. The bound $Ru(bpy)_3^{2+}$ labels' measurements are made in a TPA solution, while unbound $Ru(bpy)_3^{2+}$ labels are flushed from the cell. Some other flow cell ECL instruments were constructed on the basis of Origen design, including the NucliSens system (BioMerieux Inc.) and the PicoLumi system (Eisai Inc.) [38]. Origen bead-based instrument for the assay of Cryptosporidium parvum, a water-borne protozoan and causative agent of gastrointestinal illness in humans, is also available [39]. Roche Diagnostics Corp. is providing some new-generation ECL instrumentation with over 150 assays, including Roche ELECSYS 2010, Modular analytics E-170 systems, cobas 6,000 analyzer series, and the cobas e 411 analyzer [21].

Mesoscale discovery (MSD) Corporation has developed the ECL instruments using proprietary multi-array and multi-spot multi-well plates. These multi-well plates were integrated with disposable screen-printed carbon ink electrodes into the bottom [10]. Each well contains several binding domains (multi-spot) that react with specific targets. Biological reagents retain a high level of biological activity and can be attached to the carbon simply by passive adsorption. Only labels near the electrode are excited and detected, enabling nonwashed assays. A schematic diagram of a multi-spot plate assay for four human cytokines is shown in Fig. 3.6.

At present, MSD Corporation is providing four models of instruments, namely SECTOR Imager 6000, SECTOR Imager 2400, SECTOR PR 400, and SECTOR PR 100 (Fig. 3.7h–i). An ultralow-noise CCD camera is used in SECTOR Imager for light collection with ultimate sensitivity, wide dynamic range, and rapid read

Fig. 3.5 Schematic diagram of a flow system based on BioVeris technology using magnetic beads as a solid support for binding reactions and ECL measurements (Reprinted with permission from Ref. [10]. Copyright 2008 American Chemical Society)

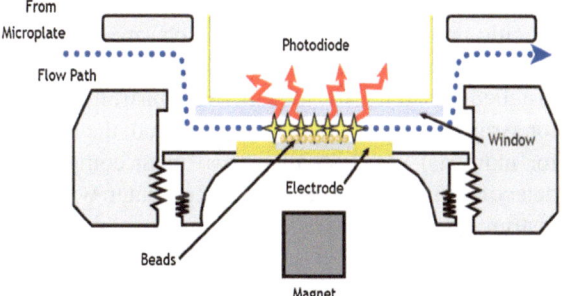

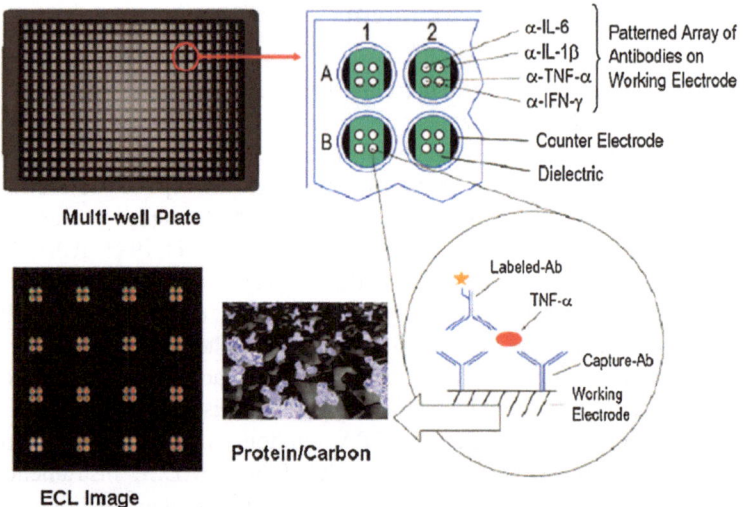

Fig. 3.6 Schematic diagram of a multi-spot plate assay for *four* human cytokines. Each spot within each well of the multi-well plate contains capture antibody specific for *one* cytokine. (Inset) Images of the ECL emitted from assays and protein immobilized on carbon electrode surface (Reprinted with permission from Ref. [10]. Copyright 2008 American Chemical Society)

times, while a photodiode array is used in Sector PR Reader for fast and efficient detection. At present, more than 150 immunoassays are available from MSD which include phosphoprotein and intracellular markers, cardiac markers, vascular markers and growth factors, fertility markers, hypoxia markers, metabolic markers, and so on [10, 36]. Wang's group at Changchun Institute of Applied Chemistry, Chinese Academy of Sciences, introduced a commercial automated CE-ECL system using PMT to detect light and is manufactured by Xi'an Remax Electronic Co. Ltd., (Xi'an, China) [40]. The system incorporates CE separation together with simultaneous EC and ECL detection, including computer-controllable data acquisition and data treatment [36].

Three models of M-series ECL analyzers are currently available from BioVeris Corp., namely M1M, M1MR, and M-384. Two of them are single-channel (M1M and M1MR), and M-384 is having eight channels. These instruments have flow-based detection system, employ an on-line automated separation to reduce non-specific interferences, and used in pharmaceutical research, industrial, academic, and government research laboratories. The M1MR analyzer is the integration of installed research software with M1M instrument and is an open system, mostly used for assay development. On the other hand, the M-384 analyzer was intended mainly for industrial use. It contains two main components: (1) a sample handling and detection unit and (2) a built-in computer with application-specific software for instrument control and data handling. Several models of commercial flow cell-based (a–g) or imaging-based (h–i) ECL instruments are shown in Fig. 3.7 [10, 41, 42].

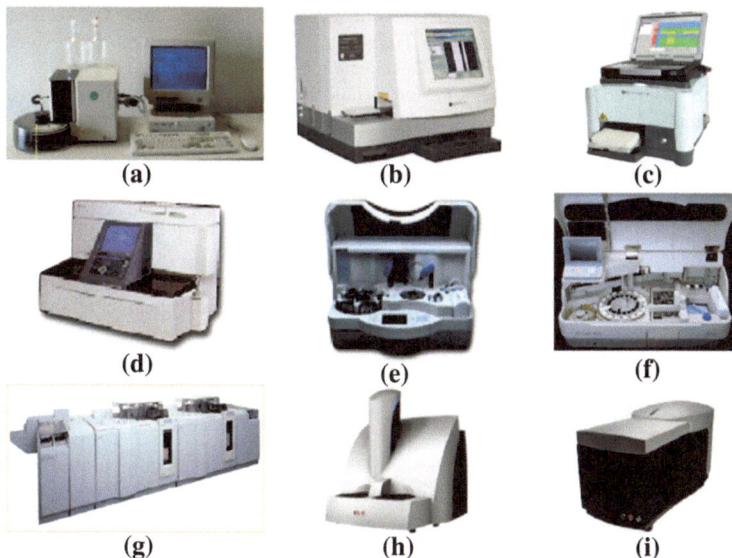

Fig. 3.7 Commercial flow cell-based (**a–g**) or imaging-based (**h–i**) ECL instrumentation: **a** ORIGEN 1.5; **b** M-series M-384 analyzers by BioVeris Corp.;66 **c** M1 M analyzers by BioVeris Corp.; **d** PicoLumi by Eisai, Japan; **e** Elecsys 1010; **f** Elecsys 2010; **g** MODULAR system containing E-170 immunoassay module by Roche Diagnostics; 66 **h** Sector Imager 6000; and **i** Sector PR Reader 400 by Meso Scale Discovery (Reprinted with permission from Ref. [10]. Copyright 2008 American Chemical Society)

Rozhitskii introduced the ELAN-2 system (Khar'kov, Ukraine) consisting of a rotating ring-disk electrode in a batch cell with sample injection facilities [43]. Bard [44] and some other groups [45, 46] described America's first ECL analyzer, the Origen-1 Analyzer (IGEN, Rockville, MD, USA). Few years back, Perkin-Elmer has also been reported to produce QPCR 5000 system for quantitative polymerase chain reaction determinations in DNA studies based on ECL [47]. Boehringer Mannheim have recently developed completely automated instruments for ECL immunoassay and DNA probe studies, the ELECSYS 10100 and ELECSYS 20100, capable of processing up to 90 samples per hour [48]. BioVeris [21] also provided a number of various immunoassays. Several immunoassays can now be performed with imaging-based ECL instruments manufactured by MSD (Fig. 3.7) [10]. Approximately 150 immunoassays are currently available from this company.

References

1. Ludvik J (2011) DC-electrochemiluminescence (ECL with a coreactant)-principle and applications in organic chemistry. J Solid State Electrochem 15(10):2065–2081. doi:10.1007/s10008-011-1546-x
2. Miao W (2007) In: Zoski CG (ed) Handbook of electrochemistry. Elsevier, The Netherlands, Amsterdam, p 541

3. Fan FRF (2004) In: Bard AJ (ed) Electrogenerated chemiluminescence. Dekker, New York
4. Fry AJ (1996) In: Kissinger PT, Heineman WR (eds) Labtoratory techniques in electroanalytical chemistry, 2nd edn. Deker, New York
5. Creager SE (2007) In: Zoski CG (ed) Handbook of electrochemistry. Elsevier, The Netherlands, Amsterdam, p 57
6. Yin XB, Dong SJ, Wang E (2004) Analytical applications of the electrochemiluminescence of tris (2.2'-bipyridyl) ruthenium and its derivatives. Trac-Trends Anal Chem 23 (6):432–s441. doi:10.1016/s0165-9936(04)00603-x
7. Du Y, Wang E (2007) Capillary electrophoresis and microchip capillary electrophoresis with electrochemical and electrochemiluminescence detection. J Sep Sci 30(6):875–890. doi:10.1002/jssc.200600472
8. Yin XB, Wang E (2005) Capillary electrophoresis coupling with electrochemilurninescence detection: a review. Anal Chim Acta 533(2):113–120. doi:10.1016/j.aca.2004.11.015
9. Danielson ND (2004) In: Bard AJ (ed) Electrogenerated chemiluminescence. Dekker, New York, p 397
10. Miao W (2008) Electrogenerated chemiluminescence and its biorelated applications. Chem Rev 108(7):2506–2553. doi:10.1021/cr068083a
11. Chi Y, Dong Y, Chen G (2007) Inhibited Ru(bpy)$_3^{2+}$ electrochemiluminescence related to electrochemical oxidation activity of inhibitors. Anal Chem 79(12):4521–4528
12. Chi Y, Duan J, Lin S, Chen G (2006) Flow injection analysis system equipped with a newly designed electrochemiluminescent detector and its application for detection of 2-thiouracil. Anal Chem 78(5):1568–1573
13. Haswell SJ (1997) Development and operating characteristics of micro flow injection based on electroosmotic flow. Analyst 122 (1):1R–10R
14. Knight AW, Greenway GM, Chesmore ED (1995) Development of a silicon photodiode, electrogenerated chemiluminescence, flow-through detector. Anal Proc Including Anal Commun 32(4):125–127
15. Collinson MM, Wightman RM (1993) High-frequency generation of electrochemiluminescence at microelectrodes. Anal Chem 65(19):2576–2582
16. Pittet P, Lu GN, Galvan JM, Ferrigno R, Blum LJ, Leca-Bouvier B (2007) PCB-based integration of electrochemiluminescence detection for microfluidic systems. Analyst 132(5):409–411
17. Hosono H, Satoh W, Fukuda J, Suzuki H (2007) On-chip handling of solutions and electrochemiluminescence detection of amino acids. Sens Actuators B: Chemical 122(2):542–548
18. Yin XB, Qiu HB, Sun XH, Yan JL, Liu JF, Wang EK (2004) Capillary electrophoresis coupled with electrochemiluminescence detection using porous etched joint. Anal Chem 76(13):3846–3850. doi:10.1021/ac049743j
19. Chiang MT, Whang CW (2001) Tris(2,2'-bipyridyl)ruthenium(III)-based electrochemiluminescence detector with indium/tin oxide working electrode for capillary electrophoresis. J Chromatogr A 934(1–2):59–66. doi:10.1016/s0021-9673(01)01279-1
20. Chiang MT, Lu MC, Whang CW (2003) A simple and low-cost electrochemiluminescence detector for capillary electrophoresis. Electrophoresis 24(17):3033–3039. doi:10.1002/elps.200305513
21. Wang X, Bobbitt DR (1999) In situ cell for electrochemically generated Ru(bpy)$_3^{3+}$-based chemiluminescence detection in capillary electrophoresis. Anal Chim Acta 383(3):213–220
22. Forbes GA, Nieman TA, Sweedler JV (1997) On-line electrogenerated Ru(bpy)$_3^{3+}$ chemiluminescent detection of I^2-blockers separated with capillary electrophoresis. Anal Chim Acta 347(3):289–293
23. Dickson JA, Ferris MM, Milofsky RE (1997) Tris (2,2'-bipyridyl)ruthenium (III) as a chemiluminescent reagent for detection in capillary electrophresis. J High Res Chromatogr 20(12):643–646
24. Huang XJ, Wang SL, Fang ZL (2002) Combination of flow injection with capillary electrophoresis: 8, Miniaturized capillary electrophoresis system with flow injection sample

introduction and electrogenerated chemiluminescence detection. Anal Chim Acta 456(2):167–175

25. Cao WD, Liu JF, Yang XR, Wang E (2002) New technique for capillary electrophoresis directly coupled with end-column electrochemiluminescence detection. Electrophoresis 23(21):3683–3691. doi:10.1002/1522-2683(200211)23

26. Du Y, Wei H, Kang JZ, Yan JL, Yin XB, Yang XR, Wang EK (2005) Microchip capillary electrophoresis with solid-state electrochemiluminescence detector. Anal Chem 77(24):7993–7997. doi:10.1021/ac051369f

27. Qiu HB, Yan JL, Sun XH, Liu JF, Cao WD, Yang XR, Wang EK (2003) Microchip capillary electrophoresis with an integrated indium tin oxide electrode-based electrochemiluminescence detector. Anal Chem 75(20):5435–5440. doi:10.1021/ac034500x

28. Preston JP, Nieman TA (1996) An electrogenerated chemiluminescence probe and its application utilizing tris(2,2′-bipyridyl)ruthenium(II) and luminol chemiluminescence without a flowing stream. Anal Chem 68(6):966–970

29. Zhan W, Alvarez J, Crooks RM (2002) Electrochemical sensing in microfluidic systems using electrogenerated chemiluminescence as a photonic reporter of redox reactions. J Am Chem Soc 124(44):13265–13270

30. Zhan W, Alvarez J, Crooks RM (2003) A two-channel microfluidic sensor that uses anodic electrogenerated chemiluminescence as a photonic reporter of cathodic redox reactions. Anal Chem 75(2):313–318. doi:10.1021/ac020488h

31. Zhan W, Alvarez J, Sun L, Crooks RM (2003) A multichannel microfluidic sensor that detects anodic redox reactions indirectly using anodic electrogenerated chemiluminescence. Anal Chem 75(6):1233–1238

32. Ding S-N, Xu J-J, Chen H-Y (2006) Microchip capillary electrophoresis coupled with an end-column electrochemiluminescence detection. Talanta 70(2):403–407

33. Ding S-N, Xu J-J, Zhang W-J, Chen H-Y (2006) Tris (2,2′-bipyridyl)ruthenium(II)-zirconia-nafion composite modified electrode applied as solid-state electrochemiluminescence detector on electrophoretic microchip for detection of pharmaceuticals of tramadol, lidocaine and ofloxacin. Talanta 70(3):572–577. doi:10.1016/j.talanta.2006.01.017

34. Yotter RA, Wilson DM (2003) A review of photodetectors for sensing light-emitting reporters in biological systems. Sensors J, IEEE 3(3):288–303

35. Zhan W, Bard AJ (2007) Electrogenerated chemiluminescence. 83, immunoassay of human C-reactive protein by using Ru(bpy)$_3^{2+}$-encapsulated liposomes as labels. Anal Chem 79(2):459–463. doi:10.1021/ac061336f

36. Hu L, Xu G (2010) Applications and trends in electrochemiluminescence. Chem Soc Rev 39(8):3275–3304. doi:10.1039/b923679c

37. Giles JH, Ridder TD, Williams RH, Jones DA, Denton MB (1998) Product review: selecting a CCD camera. Anal Chem 70(19):663A–668A

38. Rhyne PW, Wong OT, Zhang YJ, Weiner RS (2009) Electrochemiluminescence in bioanalysis. Bioanal 1(5):919–935. doi:10.4155/bio.09.80

39. Lee YM et al (2001) Development and application of a quantitative, specific assay for cryptosporidium parvum oocyst detection in high-turbidity environmental water samples. Am J Trop Med Hyg 65(1):1–9

40. Li H-J, Han S, Hu L-Z, Xu G-B (2009) Progress in Ru (bpy)$_3^{2+}$ electrogenerated chemiluminescence. Chin J Anal Chem 37(11):1557–1565

41. Li H, Shi L, Liu X, Niu W, Xu G (2009) Determination of isocyanates by capillary electrophoresis with tris(2,2′-bipyridine) ruthenium(II) electrochemiluminescence. Electrophoresis 30(22):3926–3931. doi:10.1002/elps.200900281

42. C-y Liu, Bard AJ (2009) Chemical redox reactions induced by cryptoelectrons on a PMMA surface. J Am Chem Soc 131(18):6397–6401

43. Rozhitskii NN, Kukoba AV, Belash EM, Bykh AI (1994) ELAN-2 apparatus for homogenous and heterogenous electrochemiluminescent analysis. J Anal Chem USSR 49(9):929–931

44. Carter MT, Bard AJ (1990) Electrochemical investigations of the interaction of metal chelates with DNA. 3, electrogenerated chemiluminescent investigation of the interaction of tris(1,10-phenanthroline)ruthenium(II) with DNA. Bioconjugate Chem 1(4):257–263

45. Blackburn GF, Shah HP, Kenten JH, Leland J, Kamin RA, Link J, Peterman J, Powell MJ, Shah A, Talley DB et al (1991) Electrochemiluminescence detection for development of immunoassays and DNA probe assays for clinical diagnostics. Clin Chem 37(9):1534–1539

46. Kenten JH, Casadei J, Link J, Lupold S, Willey J, Powell M, Rees A, Massey R (1991) Rapid electrochemiluminescence assays of polymerase chain reaction products. Clin Chem 37(9):1626–1632

47. Knight AW (1999) A review of recent trends in analytical applications of electrogenerated chemiluminescence. TrAC, Trends Anal Chem 18(1):47–62. doi:10.1016/S0165-9936(98)00086-7

48. Hoyle NR, Erkert B, Kraiss S (1996) Electrochemiluminescence: leading-edge technology for automated immunoassay analyte detection. Clin Chem 42(9):1576–1578

Chapter 4
ECL Luminophores

Abstract Finding new luminophores with higher ECL efficiencies and modifying a moiety of the emitter to use it for the labeling of biomolecules are the two driving forces that lead to the synthesis of a number of new ECL-emitting species in the past several years. Three categories of luminophores are discussed in this chapter, including (a) inorganic systems, which mainly contain organometallic complexes; (b) organic systems, which cover polycyclic aromatic hydrocarbons (PAHs); and (c) semiconductor nanoparticle systems.

Keywords Luminophores · Organic systems · Inorganic systems · Dendrimers · Nanoparticle systems · Nanoclusters · Carbon nanoparticles · Semiconductor nanocrystals · Quantum dots

Since the discovery of ECL process, there has been considerable advancement in ECL system. Different types of ECL systems have been introduced, such as organic, inorganic, and nanoparticle systems.

4.1 Organic Systems

The first exhaustive study on ECL involved organic systems [1–3]. In the early ages of ECL, PAHs, rubrene, and DPA have been studied extensively owing to their high fluorescence quantum yields and stable radical cations and anions in aprotic media. Some other organic systems, such as luminol [4], acridinium esters [5], polymers [6–8], and siloles [9], have also been widely explored. The reader is referred to the tremendous reviews [10–12] and ECL monograph [13] for more thorough study about organic systems. Acridinium esters, taking example of lucigenin (N,N'-dimethyl-9,9-bisacridinium) is reported ECL active [14] in the presence of hydrogen peroxide [15] and has been applied for the detection of riboflavin [16], human chorionic gonadotropin [17], and hemin [18]. The ECL reaction of methyl-9-(p-formylphenyl)acridinium carboxylate fluorosulfonate (MFPA) gives intense ECL signal and is employed as a label in bioassays [17].

Lucigenin ECL has also been extended to surfaces [19]. Self-assembled mono-layers (SAMs) modified electrodes and solutions containing Triton X-100 sur-factant molecules caused the enhancement in ECL of lucigenin.

Fluorene-substituted PAHs have been investigated with enhanced ECL effi-ciency and stability [20]. Derivatives of DPA, pyrene, and anthracene were pro-duced by using fluorene as a capping agent. These molecules have high PL quantum yields and can generate stable radical ions. On introducing the fluorene groups, steric hindrance is imparted, preventing interchromophore interactions and blocking the active position of PAH cores subject to electrochemical decompo-sition. Due to the electrochemical instability of its cationic radical, pyrene mol-ecule displays poor ECL properties. Hence, a strategy for ECL enhancement and radical stability by peripheral multidonors on pyrene derivatives has been adopted in which the ECL efficiencies of pyrene derivatives increase in proportion to the number of peripheral N, N-dimethyl aniline donors [21].

An ECL laser offers numerous advantages over traditional techniques, includ-ing tunability, lack of additional source and range of available wavelengths. Data for laser action driven by ECL have also been achieved [22]. The literature regarding photochemistry, electrochemistry, and ECL of five highly fluorescent boron-based laser dyes has been explored [23]. ECL of thiamin (vitamin B1) was described in which thiamin was used as an oxidative–reduction coreactant to generate ECL from ruthenium complexes [24]. In another strategy, rhodamine B was used as a sensitizer to enhance the weak intrinsic ECL of thiamin in alkaline solution with a very low LOD [25]. The process involves energy transfer between the excited oxidation product of thiamin and rhodamine B.

ECL of a dichromophoric molecule, 2,2′-bis (10-phenylanthracen-9-yl)-9, 9′-spirobifluorene, is reasonably explained by a simultaneous two-electron trans-fer. This dichromophoric molecule (composed of two phenylanthracenes linked by a spirobifluorene moiety) because of two redox centers go through annihilation. Di-ions are created during each potential pulse, annihilation occurs, and a single electron transfer to form an excimer is quenched by intramolecular transfer of the other electron prior to emission [26]. ECL of 3,6-dispirobifluorene-N-phenylcar-bazole which contains two spirobifluorene groups covalently attached to an N-phenylcarbazole core has also been studied [27]. Donor–acceptor architectures such as this compound may provide a general approach to design new materials exhibiting efficient ECL. ECL from non-PAHs such as perylene diimide radical anions and cations were generated by pulsing the applied potential between the parent compound's first oxidation and reductive waves [28]. Fluorene derivatives, due to their large energy gap, usually emit blue light. An electron-deficient 2,1, 3-benzothiadiazole group with a narrower energy gap after introducing into fluorine turns ECL emission color from blue to stable green. The ability to tune the luminescence of these systems suggests that this approach is promising to achieve multiple wavelength ECL labels [20]. By changing the position of 2, 5-substituents and the steric protection of both, the chromophore and the reactive parts of the substituents, an efficient and stable ECL of extended silole-based chromophores was obtained recently. By tuning the electrochemical potentials of silole–thiophene hybrid chromophores, stable radical cations favorable for ECL emission were generated [29]. The ability to tune the electrochemical potentials of these systems makes them of great interest to achieve efficient ECL materials [30].

Chen with his group members reported ECL from indole and tryptophan using hydrogen peroxide [31] with LOD 1.0×10^{-7} M for both species. Another paper explored the ECL of a heptamethine cyanine dye was reported in MeCN with TPA as coreactant [32]. Organic polymers have also gained considerable attention. Strong blue photoluminescence and ECL from $(NH_4)_2S_2O_8$-treated OH-terminated polyamidoamine (PAMAM) dendrimers were reported [33]. Such kind of blue-luminescent chemical species creates interest to gain potential applications as a new fluorophore or in aqueous ECL. Two linear, stereoregular, and structurally defined polyphenylene vinylene (PPV) derivatives were also studied. A recent example involved excimer emission from poly[distyrylbenzene-b-(ethylene oxide)]s with 12 and 16 of ethylene oxide repeating units in the backbone, respectively. The proposed mechanism for ECL involved the oxidized polymer species and the strongly reducing TPA free radical (TPA•) [34]. A semiconducting conjugated polymer poly(9,9-dioctylfluorene-co-benzothiadiazole) produced soliton-like ECL waves in the electrochemical oxidation of thin films. The waves were triggered by AuNPs embedded in the film (Fig. 4.1). The ECL "wave fronts" visualized and imaged in space and time by optical emission microscopy were observed to move freely parallel to the plane of an electrode. ECL free waves were also launched by a square-shaped scratch instead of an embedded NP. For a film with a shallow scratch, irregular patterns of ECL rather than a free wave were observed. Similarly, a dense array of triggering leaks for ultrathin films generates dendritic ECL filaments throughout the film as a result of drying cracks [35]. The proposed soliton ECL wave mechanism, which is based on lateral wave propagation, anion-gated electrochemistry, and critical polymer swelling transitions, has important implications for the design and operation of electrochemical light-emitting diodes, chemically sensitive field-effect transistors, electrochromic window coatings, and certain types of solar cells [30].

Monomer, exciplex, and excimer emission is also found in donor–acceptor functionalized luminescent hairpin peptides [36]. The ECL of 10-methylpheno-thiazine (MP) was also observed by a number of research groups [37–40]. Luminol is a classical organic species that has been widely studied and is continuously producing interest. Upon electrochemical oxidation, ECL of luminol is often generated in the presence of hydrogen peroxide in alkaline solution. Luminol ECL can sensitively detect hydrogen peroxide which may be produced, as a result of some reactions, by many biologically active substances. Consequently, luminal-ECL is thought to be widely used in the construction of various biosensors. ECL of luminol employing with various NP-modified electrodes revealed that the AuNP-modified electrodes could generate strong luminol ECL in neutral pH conditions [41–43]. The catalytic performance of AuNP-modified electrodes depended on both the size of AuNPs and the electrode substrate. These results point the development of sensitive ECL enzyme biosensors in physiological pH because of the excellent biocompatible property of AuNPs [30].

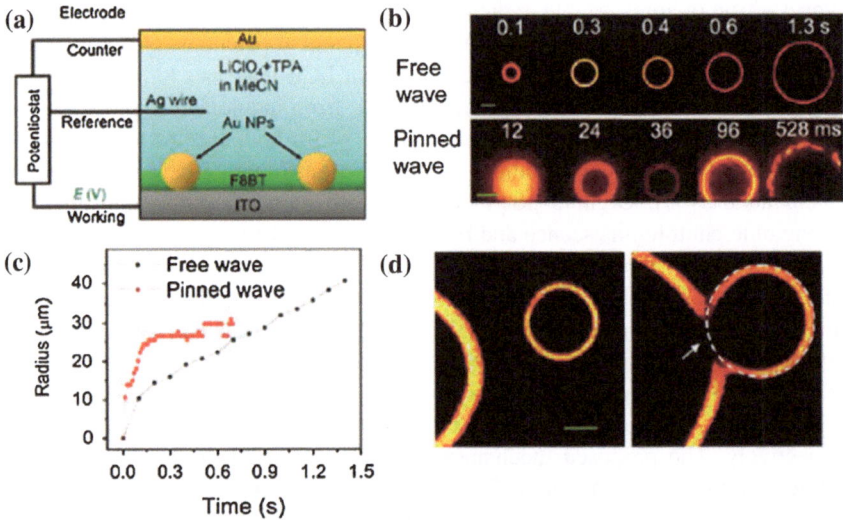

Fig. 4.1 a The electrochemical cell includes a silver-wire quasireference electrode, an Au planar counter electrode, and a 50 nm conjugated polymer-coated indium tin oxide (ITO) working electrode. **b** Images of free and pinned ECL waves triggered by 250 nm AuNPs with potential steps of 1.5 and 1.8 V, respectively. **c** Radius versus time for (**b**). **d** ECL images at (*left*) 0.3 and (*right*) 0.6 s after the potential step of 1.6 V. The integration time was 100 ms (*free*) and 12 ms (*pinned*) in (**b**) and 100 ms in (**d**). Scale bars are 20 mm. 99. Reprinted with permission from Ref. [35]. Copyright 2009 American Chemical Society

4.2 Inorganic Systems

The complexes and/or clusters containing Ag, Al, Au, Cd, Cr, Cu, Eu, Hg, Ir, Mo, W, Os, Pd, Pt, Re, Ru, Si, Tb, and Tl have been found to show ECL property. $Ru(bpy)_3^{2+}$ has been proved the first and most valuable inorganic complex to show ECL. ECL based on $Ru(bpy)_3^{2+}$ and its derivatives has many advantages of having considerably high stability and sensitivity in aqueous media with favorable electrochemical properties. $Ru(bpy)_3^{2+}$ has a characteristic of possessing compatibility with a wide range of analytes and has been used for the detection of variety of coreactants such as oxalate [44], alkylamines [45], amino acids [46], ascorbic acid [47], and many pharmaceutical compounds [48, 49]. Moreover, $Ru(bpy)_3^{2+}$ labels coupled with TPA as a coreactant has found broad applications in commercial ECL immunoassays and DNA analyses owing to offer tremendously sensitive label detection at subpicomolar concentration as well as an extremely wide dynamic range of greater than six orders of magnitude [30]. ECL of $Ru(bpy)_3^{2+}/C_2O_4^{2-}$ aqueous coreactant and $Ru(bpy)_3^{2+}$ nonaqueous (acetonitrile) annihilation by ultrasonic irradiation gives highly stable and reproducible ECL signals with less electrode fouling [50, 51].

Many methods have been adopted in order to design or modify the ligands of ruthenium complexes to improve molecular recognition ability. For instance, ruthenium complexes containing crown ether moieties covalently bonded to bipyridyl or phenanthroline ligands have been used for metal-cation sensing employing either the annihilation route or the coreactant pathway [52–57]. In few cases, the ECL intensity can be significantly enhanced up to 20-fold. Recently, with TPA as a coreactant, metal-anion ECL detection has also been published [58].

It has also been reported that strong ECL is observed from gel-entrapped $Ru(bpy)_3^{2+}$ [59], in sol–gel-derived glasses [60] and in Nafion–silica composite films [61] using coreactants. ECL of $Ru(bpy)_3^{2+}$ as a function of solution viscosity was investigated in N,N-dimethylformamide and glycerin mixtures [62]. $[Ru(bpy)_2dppz]^{2+}$ (dppz = dipyrido[3,2-a:2′,3′-c]phenazine) has ability to intercalate into DNA with high affinity due to its extensive aromatic structure [63]. It is a well-admired "light switch" molecule, as in aqueous solutions, it does not show any photoluminescence but exhibits intense photoluminescence in the presence of DNA. It also displays negligible ECL in the oxalate solution, while intercalating into the nucleic acid structure its ECL increases about 1,000 times as shown in Fig. 4.2 [64] which is attributed to the intercalation that shielded the phenazine nitrogens from the solvent and resulted in a luminescent excited state (similar to the reason of its photoluminescence properties). This dramatic increase in ECL intensity upon intercalation makes $[Ru(bpy)_2dppz]^{2+}$ a promising ECL probe for DNA interaction study and DNA-related biosensors.

Other complexes of ruthenium, such as Ru-$(bpy)_2(CE-bpy)^{2+}$ [CE-bpy is a bipyridine ligand in which a crown ether (15-crown-5) is bound to the bpy ligand in the 3- and 3′-positions] [55] and $(bpy)_2Ru(AZAbpy)^{2+}$ [(bpy) 2,2′-bipyridine; AZA-bpy) 4-(N-aza-18-crown-6-methyl-2,2′-bipyridine)] [54, 57] also exhibits strong ECL. Former chelate is sensitive to sodium ions in aqueous buffered solution (Fig. 4.3), whereas later one has been shown to be sensitive to Pb^{2+}, Hg^{2+}, Cu^{2+}, Ag^+, and K^+ in 50:50 (v/v) CH_3-CN/H_2O (0.1 M KH_2PO_4 as electrolyte) and aqueous (0.1 M KH_2PO_4) solution (Fig. 4.4).

Fig. 4.2 a Scheme of ECL switch based on $[Ru(bpy)_2dppz]^{2+}$ and DNA. **(B)** ECL intensities in 5 mM pH 5.5 oxalate solution containing 0.1 mM $[Ru(bpy)_2dppz]^{2+}$ (*curve a*) and 0.1 mM $[Ru(bpy)_2dppz]^{2+}$ + 0.16 mM DNA (*curve b*). Adapted from Ref. [64]. Copyright 2009 American Chemical Society

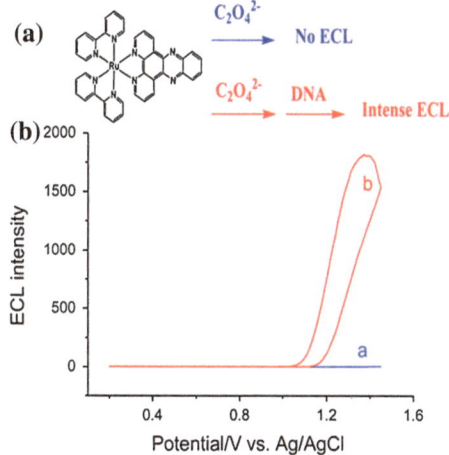

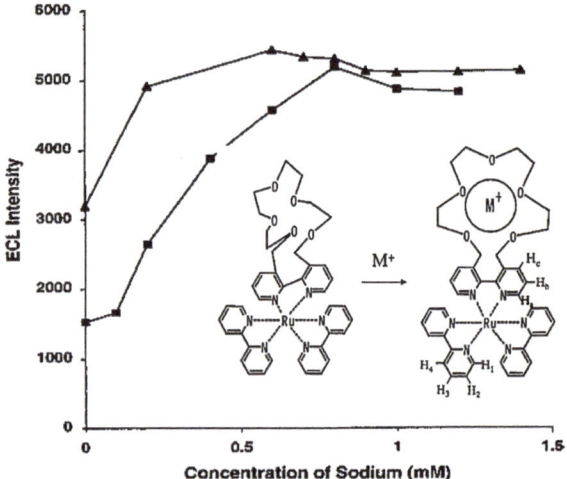

Fig. 4.3 ECL signal intensity as a function of sodium concentration for solutions containing 0.3 Ru(bpy)$_2$(CEbpy)$^{2+}$ and 30 mM TPA in 0.1 M TBAClO$_4$, MeCN (9) and in 0.1 M, pH 7.0, tris buffer (2). Adapted from Ref. [55]. Copyright 2002 American Chemical Society

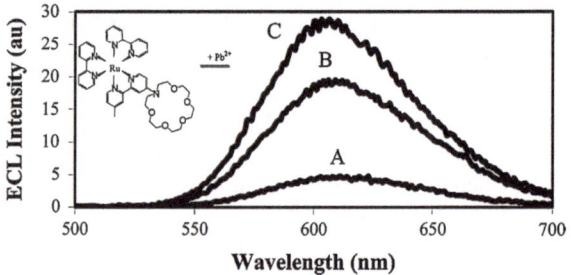

Fig. 4.4 Perturbation of ECL emission spectrum of (bpy)$_2$Ru(AZA-bpy)(PF$_6$)$_2$ (0.1 mM) in 50:50 (v/v) CH$_3$CN/H$_2$O upon addition of Pb^{2+}: (*A*) 0 mM Pb^{2+}; (*B*) 0.5 mM Pb^{2+} (5-fold excess); (*C*) 1 mM Pb^{2+} (10-fold excess). Adapted from Ref. [57]. Copyright 2002 American Chemical Society

A solution-based self-assembly approach by direct mixing of H$_2$PtCl$_6$ and Ru(bpy)$_3$Cl$_2$ aqueous solutions was employed for the synthesis of Ru(bpy)$_3$$^{2+}$-containing supramolecular microstructure [65]. These microstructures possess exceptional ECL behaviors and hence hold great promise as new luminescent materials. These structures consist of a large quantity of star-shaped microstructures with six branches developing the idea that electrostatic attractions between the positively charged Ru(bpy)$_3$$^{2+}$ and the negatively charged PtCl$_6$$^{2-}$ drive the formation of micrometer-scale supramolecular microstructures. Ru(bpz)$_3$$^{2+}$ (bpz) 2,2′-bipyrazine), another complex of ruthenium, has also attracted substantial attention and generates ECL in acetonitrile upon formation of the +3 and +1 states [66]. A characteristic of ECL of Ru(bpz)$_3$$^{2+}$ in the presence of S$_2$O$_8$$^{2-}$, as a reductive–oxidation coreactant, gives bright orange luminescence in aqueous solution [67]. This system has been used to determine persulfate with nanomolar (nM) detection limits [68]. Nanostructured materials have distinctive properties, so it is worth-interesting to observe ECL from nanoscale ruthenium complexes. ECL from single crystalline nanobelts obtained from a water-insoluble $Ru(bpy)_3^{2+}$ derivative [69], $[Ru(bpy)_2(4,4'-(CH_3(CH_2)_{14}COO)_2-bpy)](ClO_4)_2$ was easily observed using TPA as a coreactant. Moreover, ECL of a single nanobelt deposited on an ultramicroelectrode was also observed to achieve highly sensitive ECL detection.

Some other strategies, such as synthesis and behavior of multimetallic ruthenium complexes, including a trinuclear, mixed-metal supramolecular complex, $[((phen)_2-Ru(dpp))_2RhCl_2]^{5+}$ (phen = 1,10-phenanthroline, dpp = 2,3-bis(2-pyridyl)pyrazine) [70] to improve the ECL detection sensitivity, have also been investigated [71]. The first multimetallic ruthenium complex studied in ECL is the bimetallic ruthenium complex $[(bpy)_2Ru]_2(bphb)^{4+}$ (bphb = 1, 4-bis(4'-methyl-2,2'-bipyridin-4-yl)benzene) [71] which produced ECL enhancement 2–3-folds more than $Ru(bpy)_3^{2+}$ in aqueous and nonaqueous solution via annihilation method and with TPA. To multilabel biomolecules at a single site and to enhance ECL intensity with minimum impact on bioreactivity, dendritic $Ru(bpy)_3^{2+}$ labels have also been studied [72]. Three new bimetallic ruthenium complexes $[(bpy)_2Ru(bpy)(CH_2)_n(bpy)Ru(bpy)_2]^{4+}$ (n = 3, 5, 8) with different chain lengths have been synthesized and the effect of the length of carbon chain linkage were investigated by using TPA and DBAE as coreactants [73]. It was found that ECL intensities increase with increasing the length of carbon chain linkage. These results demonstrate an approach to further enhance and tune ECL efficiency by using multimetallic ruthenium complexes. Ru(II) diimine complexes having phosphonic acid substituents are adsorbed on to TiO_2-modified ITO electrodes and undergo coreactant ECL with $C_2O_4^{2-}$ [74].

Regardless of ruthenium (II) complex systems, usually oxidation is irreversible and reduction is reversible in a number of other metal polypyridine ECL systems. For these systems, commonly adopted ECL processes could not be used as this might decompose these molecules by electro-oxidation [75]. In such cases, the emission can be generated at LOP, potential that is just positive enough to oxidize TPA and is not enough to oxidize $Ru(bpy)_3^{2+}$ [76]. LOP ECL seems to provide a potential route for the generation of ECL of irreversible oxidation metal complexes. ECL of a bis-cyclometalated alkynylgold (III) (via the LOP ECL route) with irreversible oxidation but reversible reduction has been reported recently [77]. The lower potentials required in the LOP ECL also play a significant role in DNA detection because oligonucleotide sequences would be irreversibly damaged at potentials above +1.0 V versus SCE.

Despite the fact that neutral Ir(III) complexes are found to be more efficient than $Ru(bpy)_3^{2+}$, their use is limited due to their very low solubility in aqueous solutions and sensitivity to oxygen quenching [30, 78–80]. Ir-(ppy)3 (ppy) 2-phenylpyridine in acetonitrile via annihilation between the reduced, $Ir(ppy)^{3-}$, and oxidized, $Ir(ppy)^{3+}$, species displayed weak ECL [10] while in benzonitrile, strong ECL was observed [81]. ECL of F(Ir)pic [bis(3,5-difluoro-2-(2-pyri-dyl)phenyl-(2-carboxypyridyl)-iridium III] and (btp)$_2$Ir(acac) [bis[2,(2′-benzot-hienyl)pyridinato-N,C3′](acetylacetonate)-Ir(III)] in acetonitrile (MeCN), mixed MeCN/H$_2$O (50:50, v/v), and aqueous solutions using TPA has also been inves-tigated [80]. In the presence of surfactant, ECL enhancement of up to 6-fold and 20-fold for F(Ir)pic and (btp)$_2$Ir(acac) has been observed, respectively.

Stable ECL of a completely insoluble neutral Ir(III) complex Ir(pq)$_2$acac in aqueous media has recently been studied by encapsulating it into silica-PEG NPs [10], thus, making the ECL detection feasible under typical bioassay conditions by improving solubilization, limiting water and oxygen quenching which offers a promising way to obtain a great number of water soluble ECL materials. Three ionic iridium complexes such as, [Ir(ppy)$_2$(bpy)]Cl$_2$ (ppy = 2-phenylpyridine), [Ir(ppy)$_2$(phen)]Cl$_2$, and [Ir(ppy)$_2$(BPS)]Na (BPS = 4,7-diphenyl-1,10-phenanth-rolinedisulfonate) have recently been investigated which showed favorable solu-bility in aqueous solutions [82]. The obtained results showed that ionic iridium complexes hold great promise for ECL and chemiluminescent detection in aqueous solutions. Some more detailed literature on inorganic systems may be found in these excellent reviews [10–12].

A thin-layer cell composed of glass/ITO/emitting solution/ITO/glass was used in this experiment. Another paper reported the role of triplet excited state on the annihilation ECL of Ir(ppy)$_3$ [83] and the annihilation/coreactant ECL of Ir(ppy)$_3$ in acetonitrile/dioxane (1:1 v/v) solutions containing 0.1 M TBAPF$_6$ as the elec-trolyte was also reported [84]. Al(III) has been shown to play a role in some disorders in human body, e.g., bone, neurological disorders, and Alzheimer's disease [85]. Therefore, there is a great need to detect Al(III) in medical, clinical, and environmental samples; especially using ECL, it would be more advantageous. One of the aluminum complexes exhibiting ECL properties, synthesized in aqueous, buffered solutions is tris(8-hydroxyquinoline-5-sulfonic acid)alumi-num(III) [86]. The ECL of Al(HQS)$_3$ was measured using TPA in aqueous buf-fered solution. ECL has also been extended to copper bis(phenanthroline) systems [62, 87]. For example, in situ ECL generation from aqueous solutions of Cu(dmp)$^{2+}$ from copper ions and the ligand 9,10-dimethylphenanthroline (dmp) was also reported [87]. Reduction of Cu^{2+} ions (the common oxidation state found in nature) to Cu$^+$ with hydroxylamine hydrochloride followed by complexation with dmp and oxidation in the presence of tri-n-propylamine is the process of ECL generation.

ECL of Ir(ppy)$_3$/TPA [79] and Os(phen)$_2$(dppene)$^{2+}$/TPA [88] in the presence of Triton X-100 were also studied. An osmium complex, Os(phen)$_3^{2+}$, was the first osmium complex that gave ECL response using S$_2$O$_8^{2-}$ reduction to generate the excited state [89]. A series of osmium complexes containing bpy and phen ligands showed ECL enhancement via annihilation [10]. While using TPA as coreactant [90], Os(phen)$_2$(dppene)$^{2+}$ exhibits ECL in aqueous and mixed aqueous/ nonaqueous solutions. Osmium polypyridine complexes have also been studied [88, 89].

4.3 Nanoparticle Systems

Nanoparticles were first of all studied in 2002 in the field of electrogenerated chemiluminescence. ECL from semiconductor Si NPs was produced in MeCN solutions via annihilation and coreactant pathway [91]. Semiconductor nanocrystals, also known as quantum dots (QDs), have also been widely studied. Few years back, ECL from CdSe [92], CdS [93], CdTe [53], and ZnS [94], PbS and ZnSe QDs [95, 96] was observed which displayed applications in the field of biosensors.

In addition to these compound semiconductors (e.g., CdS, CdSe, and CdTe), elemental semiconductors (e.g., Si and Ge), many can also produce ECL. The ECL mechanism of semiconductor NPs follows the general annihilation and coreactant ECL reaction pathways. Many other NPs such as graphene oxide NPs [97], semiconductor NPs [98–100], and carbon NPs [101] can also produce ECL. For example, ECL from a suspension of graphene oxide NPs was detected by using a coreactant at relatively high concentrations. Besides suspension of graphene oxide NPs, an electrochemically oxidized graphite layer on highly oriented pyrolytic graphite has been reported to generate ECL. Usually, hydrogen peroxide or peroxydisulfate are used as coreactants at high potentials to generate semiconductor NP ECL. While in recent years amines such as DBAE are found to be used as coreactant with semiconductor NPs to generate strong ECL at low potential [102]. Furthermore, it was also discovered that DBAE, a tertiary amine, is the most effective one. ECL study of CdSe/ZnSe core/shell type NPs was reported by Bard group [103]. CdSe NPs that were well-passivated with a shell of ZnSe showed a large ECL peak at the wavelength of band-edge PL plus a red shift by ~ 200 nm from the PL peak. NP films including PbSe [104], CdSe thin film and single monolayers of CdSe in molecular organic devices [105], CdSe/ZnS [106], ZnS/CdSe, and CdSe/CdS core/shell type films were also reported [107]. Studies relating these systems are important because NPs have found some applications in optoelectronic systems or as components in future nanoelectronic devices.

Recently, carbon NPs with ECL activity have been prepared by an electrochemical method [101] as well as through a facile and economical microwave pyrolysis approach [108]. On heating, a saccharide and surfactant PEG-200 mixture solution in a 500 W microwave oven for 2–10 min, the colorless solution was transformed to yellow and finally to dark brown, which entails the formation of carbon NPs. This case represents a convenient and low-cost method which decreases the reaction time and devotes a potential advancement to large-scale industrialization. It's highly sensitive analytical applications emerged by the advantages of convenience, low cost, and effectiveness of the method.

Silver nanoclusters exhibiting cathodic hot electron-induced ECL with considerable potential of its use as probes in molecular sensing have also been reported [109]. Most of the ECL studies have been performed using organic molecules; however, few studies have been performed on organic NP ECL [110]. By direct reaction between $HAuCl_4$ and luminol in aqueous solutions, luminol-reduced AuNPs have been synthesized, in recent years, where diameter of the NPs increases with decrease in luminal concentration. The presence of cysteine may enhance the ECL of luminol-capped AuNPs by 20-fold [111, 112]. The biocompatibility of Au cores and the relative strong ECL intensity make the luminol-capped AuNPs potential ECL biomarkers for their applications in biosensors [30]. Sterically stabilized silicon NCs are also reported to have ECL property which is attributed to the electron transfer reactions between positively and negatively charged NCs (i.e., annihilation) or between charged NCs and coreactants results in ECL [113]. ECL from trioctylphosphineoxide (TOPO)-capped CdSe nanocrystals dissolved in methylene chloride containing 0.1 M tetra-n-butylammonium perchlorate (TBAClO$_4$) is also reported [92]. For a detailed study on the electrochemical and ECL behaviors of semiconductor NPs in solutions and in films, a comprehensive and excellent draft [114] of this field is present to go through.

References

1. Hercules DM (1964) Chemiluminescence resulting from electrochemically generated species. Science 145(3634):808–809. doi:10.1126/science.145.3634.808 (New York, NY)
2. Santhanam KSV, Bard AJ (1965) Chemiluminescence of electrogenerated 9,10-diphenylanthracene anion radical. J Am Chem Soc 87(1):139–140. doi:10.1021/ja01079a039
3. Visco RE, Chandross EA (1964) Electroluminescence in solutions of aromatic hydrocarbons. J Am Chem Soc 86(23):5350–5351
4. Leca B, Blum LJ (2000) Luminol electrochemiluminescence with screen-printed electrodes for low-cost disposable oxidase-based optical sensors. Analyst 125(5):789–791
5. Littig JS, Nieman TA (1992) Quantitation of acridinium esters using electrogenerated chemiluminescence and flow injection. Anal Chem 64(10):1140–1144
6. Fan FRF, Mau A, Bard AJ (1985) Electrogenerated chemiluminescence, a chemiluminescent polymer based on poly(vinyl-9,10-diphenylanthracene). Chem Phy Lett 116(5):400–404
7. Prieto I, Teetsov J, Fox MA, Vanden Bout DA, Bard AJ (2000) A study of excimer emission in solutions of poly(9,9-dioctylfluorene) using electrogenerated chemiluminescence. J Phy Chem A 105(3):520–523
8. Richter MM, Fan F-RF, Klavetter F, Heeger AJ, Bard AJ (1994) Electrochemistry and electrogenerated chemiluminescence of films of the conjugated polymer 4-methoxy-(2-ethylhexoxyl)-2,5-polyphenylenevinylene. Chem Phy Lett 226(12):115–120
9. Sartin MM, Boydston AJ, Pagenkopf BL, Bard AJ (2006) Electrochemistry, spectroscopy, and electrogenerated chemiluminescence of silole-based chromophores. J Am Chem Soc 128(31):10163–10170
10. Richter MM (2004) Electrochemiluminescence (ECL). Chem Rev 104(6):3003–3036. doi:10.1021/cr020373d
11. Pyati R, Richter MM (2007) ECL-electrochemical luminescence. Ann Rep Sec C (Phyl Chem) 103(0):12–78
12. Miao W (2008) Electrogenerated chemiluminescence and its biorelated applications. Chem Rev 108(7):2506–2553. doi:10.1021/cr068083a
13. Bard AJ (2004) Electrogenerated chemiluminescence. Taylor & Francis, London
14. Legg KD, Hercules DM (1969) Electrochemically generated chemiluminescence of lucigenin. J Am Chem Soc 91(8):1902–1907

15. Haapakka KE, Kankare JJ (1981) Electrogenerated chemiluminescence of lucigenin in aqueous alkaline solutions at a platinum electrode. Anal Chim Acta 130(2):415–418

16. Zhang C, Qi H (2002) Highly sensitive determination of riboflavin based on the enhanced electrogenerated chemiluminescence of lucigenin at a platinum electrode in a neutral aqueous solution. Anal Sci 18(7):819–822. doi:10.2116/analsci.18.819

17. Lin JM, Yamada M (1998) Electrogenerated chemiluminescence of methyl-9-(p-formylphenyl) acridinium carboxylate fluorosulfonate and its applications to immunoassay. Microchem J 58(1):105–116. doi:10.1006/mchj.1997.1539

18. Nan Chen G, Zhang L, Er Lin R, Cong Yang Z, Ping Duan J, Qing Chen H, Brynn Hibbert D (2000) The electrogenerated chemiluminescent behavior of hemin and its catalytic activity for the electrogenerated chemiluminescence of lucigenin. Talanta 50(6):1275–1281

19. Okajima T, Ohsaka T (2002) Electrogenerated chemiluminescence of lucigenin enhanced by the modifications of electrodes with self-assembled monolayers and of solutions with surfactants. J Electroanal Chem 534(2):181–187

20. Omer KM, Ku S-Y, Wong K-T, Bard AJ (2009) Efficient and stable blue electrogenerated chemiluminescence of fluorene-substituted aromatic hydrocarbons. Angew Chem Int Ed 48(49):9300–9303

21. Oh JW, Lee YO, Kim TH, Ko KC, Lee JY, Kim H, Kim JS (2009) Enhancement of electrogenerated chemiluminescence and radical stability by peripheral multidonors on alkynylpyrene derivatives. Angew Chem Int Ed 48(14):2522–2524

22. Horiuchi T, Niwa O, Hatakenaka N (1998) Evidence for laser action driven by electrochemiluminescence. Nature 394(6694):659–661

23. Lai RY, Bard AJ (2003) Electrogenerated chemiluminescence 71. Photophysical, electrochemical, and electrogenerated chemiluminescent properties of selected dipyrromethene BF2 dyes. J Phy Chem B 107(21):5036–5042

24. Chen X, Chen W, Wang XR (2000) Electrochemiluminescent behaviour of vitamin B-1 (thiamine) and tris(2,2′-bipyridyl) ruthenium in a flow system. Acta Chim Sinica 58(5):563–566

25. Zhang C, Zhou G, Zhang Z, Aizawa M (1999) Highly sensitive electrochemical luminescence determination of thiamine. Anal Chim Acta 394(23):165–170

26. Sartin MM, Shu C, Bard AJ (2008) Electrogenerated chemiluminescence of a spirobifluorene-linked bisanthracene: a possible simultaneous, two-electron transfer. J Am Chem Soc 130(15):5354–5360

27. Rashidnadimi S, Hung TH, Wong KT, Bard AJ (2007) Electrochemistry and electrogenerated chemiluminescence of 3,6-Di(spirobifluorene)-N-phenylcarbazole. J Am Chem Soc 130(2):634–639

28. Lee SK, Zu Y, Herrmann A, Geerts Y, MÃllen K, Bard AJ (1999) Electrochemistry, spectroscopy and electrogenerated chemiluminescence of perylene, terrylene, and quaterrylene diimides in aprotic solution. J Am Chem Soc 121(14):3513–3520

29. Booker C, Wang X, Haroun S, Zhou J, Jennings M, Pagenkopf BL, Ding Z (2008) Tuning of electrogenerated silole chemiluminescence. Angew Chem Int Ed 47(40):7731–7735

30. Hu L, Xu G (2010) Applications and trends in electrochemiluminescence. Chem Soc Rev 39(8):3275–3304. doi:10.1039/b923679c

31. Chen GN, Lin RE, Zhao ZF, Duan JP, Zhang L (1997) Electrogenerated chemiluminescence for determination of indole and tryptophan. Anal Chim Acta 341(23):251–256

32. Lee SK, Richter MM, Strekowski L, Bard AJ (1997) Electrogenerated chemiluminescence. 61. Near-IR electrogenerated chemiluminescence, electrochemistry, and spectroscopic properties of a heptamethine cyanine dye in MeCN. Anal Chem 69(20):4126–4133

33. Lee WI, Bae Y, Bard AJ (2004) Strong blue photoluminescence and ECL from OH-terminated PAMAM dendrimers in the absence of gold nanoparticles. J Am Chem Soc 126(27):8358–8359

34. Rosado DJ, Miao W, Sun Q, Deng Y (2006) Electrochemistry and electrogenerated chemiluminescence of all-trans conjugated polymer poly[distyrylbenzene-b-(ethylene oxide)]s. J Phy Chem B 110(32):15719–15723

35. Chang YL, Palacios RE, Chen JT, Stevenson KJ, Guo S, Lackowski WM, Barbara PF (2009) Electrogenerated chemiluminescence of soliton waves in conjugated polymers. J Am Chem Soc 131(40):14166–14167

36. Strauß J, Daub J (2002) Donor–acceptor functionalized luminescent hairpin peptides: electrochemiluminescence of pyrene/phenothiazine-substituted optically active systems. Adv Mat 14(22):1652–1655

37. Bezman R, Faulkner LR (1972) Mechanisms of chemiluminescent electron-transfer reactions. VI. Absolute measurements of luminescence from the fluoranthene-10-methylphenothiazine system in N, N-dimethylformamide. J Am Chem Soc 94(18):6331–6337

38. Faulkner LR, Freed DJ (1971) Mechanisms of chemiluminescent electron-transfer reactions. II. Triplet yield of electron transfer in the fluoranthene-10-methylphenothiazine system. J Am Chem Soc 93(15):3565–3568

39. Faulkner LR, Freed DJ (1971) Mechanisms of chemiluminescent electron-transfer reactions. I. Role of the triplet state in energy-deficient systems. J Am Chem Soc 93(9):2097–2102

40. Slaterbeck AF, Meehan TD, Gross EM, Wightman RM (2002) Selective population of excited states during electrogenerated chemiluminescence with 10-methylphenothiazine. J Phy Chem B 106(23):6088–6095

41. Dong YP, Cui H, Wang CM (2006) Electrogenerated chemiluminescence of luminol on a gold-nanorod-modified gold electrode. J Phy Chem B 110(37):18408–18414. doi:10.1021/jp062396s

42. Dong Y-P, Cui H, Xu Y (2006) Comparative studies on electrogenerated chemiluminescence of luminol on gold nanoparticle modified electrodes. Langmuir 23(2):523–529

43. Cui H, Xu Y, Zhang Z-F (2004) Multichannel electrochemiluminescence of luminol in neutral and alkaline aqueous solutions on a gold nanoparticle self-assembled electrode. Anal Chem 76(14):4002–4010

44. Chang M–M, Saji T, Bard AJ (1977) Electrogenerated chemiluminescence. 30. Electrochemical oxidation of oxalate ion in the presence of luminescers in acetonitrile solutions. J Am Chem Soc 99(16):5399–5403. doi:10.1021/ja00458a028

45. Zu Y, Bard AJ (2000) Electrogenerated chemiluminescence. 66. The role of direct coreactant oxidation in the ruthenium tris(2,2′)bipyridyl/tripropylamine system and the effect of halide ions on the emission intensity. Anal Chem 72(14):3223–3232

46. Skotty DR, Lee WY, Nieman TA (1996) Determination of dansyl amino acids and oxalate by HPLC with electrogenerated chemiluminescence detection using tris(2,2′-bipyridyl)-ruthenium(II) in the mobile phase. Anal Chem 68(9):1530–1535. doi:101021/ac951087n

47. Zorzi M, Pastore P, Magno F (2000) A single calibration graph for the direct determination of ascorbic and dehydroascorbic acids by electrogenerated luminescence based on Ru(bpy)$_3^{2+}$ in aqueous solution. Anal Chem 72(20):4934–4939. doi:10.1021/ac991222m

48. Xu G, Dong S (2000) Electrochemiluminescent detection of chlorpromazine by selective preconcentration at a lauric acid-modified carbon paste electrode using tris(2,2′-bipyridine)ruthenium(II). Anal Chem 72(21):5308–5312

49. Shi Liu, Li Xu (2006) Electrochemiluminescent detection based on solid-phase extraction at tris(2,2′-bipyridyl)ruthenium(II)-modified ceramic carbon electrode. Anal Chem 78(20):7330–7334

50. Malins C, Vandeloise R, Walton D, Vander Donckt E (1997) Ultrasonic modification of light emission from electrochemiluminescence processes. J Phy Chem A 101(28):5063–5068

51. Walton DJ, Phull SS, Bates DM, Lorimer JP, Mason TJ (1992) Sonochemical enhancement of electrochemiluminescence. Ultrasonics 30(3):186–191

52. Schmittel M, Lin H-W (2007) Quadruple-channel sensing: a molecular sensor with a single type of receptor site for selective and quantitative multi-ion analysis. Angew Chem Int Ed 46(6):893–896

53. Bae Y, Myung N, Bard AJ (2004) Electrochemistry and electrogenerated chemiluminescence of cdTe nanoparticles. Nano Lett 4(6):1153–1161

54. Bruce D, Richter MM (2002) Electrochemiluminescence in aqueous solution of a ruthenium(ii) bipyridyl complex containing a crown ether moiety in the presence of metal ions. Analyst 127(11):1492–1494

55. Lai RY, Chiba M, Kitamura N, Bard AJ (2001) Electrogenerated chemiluminescence. 68. Detection of sodium ion with a ruthenium(ii) complex with crown ether moiety at the 3,3′-positions on the 2,2′-bipyridine ligand. Anal Chem 74(3):551–553

56. Chen Y, Lin Z, Chen J, Sun J, Zhang L, Chen G (2007) New capillary electrophoresis-electrochemiluminescence detection system equipped with an electrically heated $Ru(bpy)_3^{2+}$/multi-wall-carbon-nanotube paste electrode. J Chromatogr A 1172(1):84–91 Epub 2007 Oct 22

57. Muegge BD, Richter MM (2001) Electrochemiluminescent detection of metal cations using a ruthenium(ii) bipyridyl complex containing a crown ether moiety. Anal Chem 74(3):547–550

58. Berni E, Gosse I, Badocco D, Pastore P, Sojic N, Pinet S (2009) Differential photoluminescent and electrochemiluminescent detection of anions with a modified ruthenium(ii)–bipyridyl complex. Chem A Eur J 15(20):5145–5152

59. Collinson MM, Taussig J, Martin SA (1999) Solid-state electrogenerated chemiluminescence from gel-entrapped ruthenium(ii) tris(bipyridine) and tripropylamine. Chem Mat 11(9):2594–2599

60. Collinson MM, Novak B, Martin SA, Taussig JS (2000) Electrochemiluminescence of ruthenium(II) tris(bipyridine) encapsulated in sol gel glasses. Anal Chem 72(13):2914–2918

61. Khramov AN, Collinson MM (2000) Electrogenerated chemiluminescence of tris(2,2′-bipyridyl)ruthenium(II) ion-exchanged in nafion silica composite films. Anal Chem 72(13):2943–2948

62. McCall J, Bruce D, Workman S, Cole C, Richter MM (2001) Electrochemiluminescence of copper(I) Bis(2,9-dimethyl-1,10-phenanthroline). Anal Chem 73(19):4617–4620

63. Friedman AE, Chambron JC, Sauvage JP, Turro NJ, Barton JK (1990) A molecular light switch for DNA: $Ru(bpy)_2(dppz)^{2+}$. J Am Chem Soc 112(12):4960–4962

64. Hu L, Bian Z, Li H, Han S, Yuan Y, Gao L, Xu G (2009) $[Ru(bpy)2d\ ppz]^{2+}$ electrochemiluminescence switch and its applications for DNA interaction study and label-free ATP aptasensor. Anal Chem 81(23):9807–9811

65. Sun X, Du Y, Zhang L, Dong S, Wang E (2007) Luminescent supramolecular microstructures containing $Ru(bpy)_3^{2+}$: solution-based self-assembly preparation and solid-state electrochemiluminescence detection application. Anal Chem 79(6):2588–2592. doi:10.1021/ac062130h

66. Gonzales-Velasco J, Rubinstein I, Crutchley RJ, Lever ABP, Bard AJ (1983) Electrogenerated chemiluminescence. 42. Electrochemistry and electrogenerated chemiluminescence of the tris(2,2′-bipyrazine)ruthenium(II) system. Inorg Chem 22(5):822–825

67. Mark R (2004) Metal chelate systems. Electrogenerated chemiluminescence. CRC Press, Boca Raton, pp 301–358

68. Knight AW, Greenway GM (1994) Occurrence, mechanisms and analytical applications of electrogenerated chemiluminescence—review. Analyst 119(5):879–890. doi:10.1039/an9941900879

69. Yu J, Fan F-RF, Pan S, Lynch VM, Omer KM, Bard AJ (2008) Spontaneous formation and electrogenerated chemiluminescence of tris(bipyridine) Ru(II) derivative nanobelts. J Am Chem Soc 130(23):7196–7197

70. Wang S, Milam J, Ohlin AC, Rambaran VH, Clark E, Ward W, Seymour L, Casey WH, Holder AA, Miao W (2009) Electrochemical and electrogenerated chemiluminescent studies of a trinuclear complex, $[((phen)_2Ru(dpp))_2RhCl_2]^{5+}$, and its interactions with calf thymus DNA. Anal Chem 81(10):4068–4075. doi:10.1021/ac900282y

71. Richter MM, Bard AJ, Kim W, Schmehl RH (1998) Electrogenerated chemiluminescence. 62. Enhanced ECL in bimetallic assemblies with ligands that bridge isolated chromophores. Anal Chem 70(2):310–318. doi:10.1021/ac970736n

72. Zhou M, Roovers J, Robertson GP, Grover CP (2003) Multilabeling biomolecules at a single site. 1. Synthesis and characterization of a dendritic label for electrochemiluminescence assays. Anal Chem 75(23):6708–6717
73. Sun S, Yang Y, Liu F, Pang Y, Fan J, Sun L, Peng X (2009) Study of highly efficient bimetallic ruthenium tris-bipyridyl ecl labels for coreactant system. Anal Chem 81(24):10227–10231
74. Andersson A-M, Isovitsch R, Miranda D, Wadhwa S, Schmehl RH (2000) Electrogenerated chemiluminescence from Ru bipyridylphosphonic acid complexes adsorbed to mesoporous TiO/ITO electrodes. Chem Comm 0(6):505–506
75. Lo KKW, Chung CK, Lee TKM, Lui LH, Tsang KHK, Zhu N (2003) New luminescent cyclometalated iridium(iii) diimine complexes as biological labeling reagents. Inorg Chem 42(21):6886–6897
76. Miao W, Choi JP, Bard AJ (2002) Electrogenerated chemiluminescence 69: the tris(2,2'-bipyridine)ruthenium(II), $(Ru(bpy)_3^{2+}$/tri-n-propylamine (TPrA) system revisited-a new route involving TPrA* + cation radicals. J Am Chem Soc 124(48):14478–14485
77. Chen Z, Wong KMC, Au VKM, Zu Y, Yam VWW (2009) Electrogenerated chemiluminescence of a bis-cyclometalated alkynylgold(iii) complex with irreversible oxidation using tri-n-propylamine as co-reactant. Chem Comm 0(7):791–793
78. Bruce D, Richter MM (2002) Green electrochemiluminescence from ortho-metalated tris (2-phenylpyridine)iridium(III). Anal Chem 74(6):1340–1342
79. Cole C, Muegge BD, Richter MM (2003) Effects of poly(ethylene glycol) tert-octylphenyl ether on tris(2-phenylpyridine)iridium(III) tripropylamine electrochemiluminescence. Anal Chem 75(3):601–604
80. Muegge BD, Richter MM (2003) Multicolored electrogenerated chemiluminescence from ortho-metalated iridium(iii) systems. Anal Chem 76(1):73–77
81. Nishimura K, Hamada Y, Tsujioka T, Shibata K, Fuyuki T (2001) Solution electrochemiluminescent cell using tris(phenylpyridine) iridium. Jap J App Phy 40 (Part 2, No. 9A/B):L945–L947. doi:10.1143/JJAP.40.L945
82. Kiran RV, Zammit EM, Hogan CF, James BD, Barnett NW, Francis PS (2009) Chemiluminescence from reactions with bis-cyclometalated iridium complexes in acidic aqueous solution. Analyst 134(7):1297–1298
83. Gross EM, Armstrong NR, Wightman RM (2002) Electrogenerated chemiluminescence from phosphorescent molecules used in organic light-emitting diodes. J Electrochem Soc 149(5):E137–E142. doi:10.1149/1.1464137
84. Kapturkiewicz A, Angulo G (2003) Extremely efficient electrochemiluminescence systems based on tris(2-phenylpyridine)iridium(iii). Dalton Trans 0(20):3907–3913
85. Martin RB (1994) Aluminum: a neurotoxic product of acid rain. Acc Chem Res 27(7):204–210
86. Muegge BD, Brooks S, Richter MM (2003) Electrochemiluminescence of tris(8-hydroxyquinoline-5-sulfonic acid)aluminum(III) in aqueous solution. Anal Chem 75(5):1102–1105
87. High B, Bruce D, Richter MM (2001) Determining copper ions in water using electrochemiluminescence. Anal Chim Acta 449(12):17–22
88. Walworth J, Brewer KJ, Richter MM (2004) Enhanced electrochemiluminescence from $Os(phen)_2(dppene)^{2+}$ (phen = 1,10-phenanthroline and dppene = bis(diphenylphosphino)-ethene) in the presence of triton X-100 (polyethylene glycol tert-octylphenyl ether). Anal Chim Acta 503(2):241–245
89. Bolletta F, Rossi A, Balzani V (1981) Chemiluminescence on oxidation of tris(2,2'-bipyridine)chromium(II): chemical generation of a metal centered excited state. Inorg Chim Acta 53:L23–L24
90. Bruce D, Richter MM, Brewer KJ (2002) Electrochemiluminescence from $Os(phen)_2(dppene)^{2+}$ (phen = 1,10-phenanthroline and dppene = bis(diphenylphosphino)-ethene). Anal Chem 74(13):3157–3159

91. Ding Z, Quinn BM, Haram SK, Pell LE, Korgel BA, Bard AJ (2002) Electrochemistry and electrogenerated chemiluminescence from silicon nanocrystal quantum dots. Science 296(5571):1293–1297

92. Myung N, Ding Z, Bard AJ (2002) Electrogenerated chemiluminescence of CdSe nanocrystals. Nano Lett 2(11):1315–1319

93. Haram SK, Quinn BM, Bard AJ (2004) Nano Lett 4:183–185

94. Shen L, Cui X, Qi H, Zhang C (2007) Electrogenerated chemiluminescence of ZnS nanoparticles in alkaline aqueous solution. J Phy Chem C 111(23):8172–8175

95. Sun L, Bao L, Hyun B-R, Bartnik AC, Zhong Y-W, Reed JC, Pang D-W, Abrun HD, Malliaras GG, Wise FW (2008) Electrogenerated chemiluminescence from PbS quantum dots. Nano Lett 9(2):789–793

96. Hu X, Han H, Hua L, Sheng Z (2010) Electrogenerated chemiluminescence of blue emitting ZnSe quantum dots and its biosensing for hydrogen peroxide. Biosens Bioelectron 25(7):1843–1846

97. Fan F-RF, Park S, Zhu Y, Ruoff RS, Bard AJ (2008) Electrogenerated chemiluminescence of partially oxidized highly oriented pyrolytic graphite surfaces and of graphene oxide nanoparticles. J Am Chem Soc 131(3):937–939

98. Liu X, Jiang H, Lei J, Ju H (2007) Anodic electrochemiluminescence of CdTe quantum dots and its energy transfer for detection of catechol derivatives. Anal Chem 79(21):8055–8060. doi:10.1021/ac070927i

99. Jie G, Liu B, Pan H, Zhu J-J, Chen H-Y (2007) CdS nanocrystal-based electrochemiluminescence biosensor for the detection of low-density lipoprotein by increasing sensitivity with gold nanoparticle amplification. Anal Chem 79(15):5574–5581

100. Wang Y, Lu J, Tang L, Chang H, Li J (2009) Graphene oxide amplified electrogenerated chemiluminescence of quantum dots and its selective sensing for glutathione from thiol-containing compounds. Anal Chem 81(23):9710–9715

101. Zheng L, Chi Y, Dong Y, Lin J, Wang B (2009) Electrochemiluminescence of water-soluble carbon nanocrystals released electrochemically from graphite. J Am Chem Soc 131(13):4564–4565

102. Zhang L, Zou X, Ying E, Dong S (2008) Quantum dot electrochemiluminescence in aqueous solution at lower potential and its sensing application. J Phy Chem C 112(12):4451–4454

103. Myung N, Bae Y, Bard AJ (2003) Effect of surface passivation on the electrogenerated chemiluminescence of CdSe/ZnSe nanocrystals. Nano Lett 3(8):1053–1055

104. Wehrenberg BL, Guyot-Sionnest P (2003) Electron and hole injection in PbSe quantum dot films. J Am Chem Soc 125(26):7806–7807

105. Guyot-Sionnest P, Wang C (2003) Fast voltammetric and electrochromic response of semiconductor nanocrystal thin films. J Phy Chem B 107(30):7355–7359

106. Hikmet RAM, Talapin DV, Weller H (2003) Study of conduction mechanism and electroluminescence in CdSe/ZnS quantum dot composites. J App Phys 93(6):3509–3514. doi:10.1063/1.1542940

107. Poznyak SK, Talapin DV, Shevchenko EV, Weller H (2004) Quantum dot chemiluminescence. Nano Lett 4(4):693–698

108. Zhu H, Wang X, Li Y, Wang Z, Yang F, Yang X (2009) Microwave synthesis of fluorescent carbon nanoparticles with electrochemiluminescence properties. Chem Comm 0(34):5118–5120

109. Díez I, Pusa M, Kulmala S, Jiang H, Walther A, Goldmann AS, Müller AHE, Ikkala O, Ras RHA (2009) Color tunability and electrochemiluminescence of silver nanoclusters. Angew Chem Int Ed 48(12):2122–2125

110. Omer KM, Bard AJ (2009) Electrogenerated chemiluminescence of aromatic hydrocarbon nanoparticles in an aqueous solution. J Phy Chem C 113(27):11575–11578

111. Cui H, Wang W, Duan C-F, Dong Y-P, Guo J-Z (2007) Synthesis, characterization, and electrochemiluminescence of luminol-reduced gold nanoparticles and their application in a hydrogen peroxide sensor. Chem Eur J 13(24):6975–6984. doi:10.1002/chem.200700011

112. Wang W, Xiong T, Cui H (2008) Fluorescence and electrochemiluminescence of luminol-reduced gold nanoparticles: photostability and platform effect. Langmuir 24(6):2826–2833
113. Ding Z, Quinn BM, Haram SK, Pell LE, Korgel BA, Bard AJ (2002) Electrochemistry and electrogenerated chemiluminescence from silicon nanocrystal quantum dots. Science 296(5571):1293–1297. doi:10.1126/science.1069336 (New York, NY)
114. Bard AJ, Ding Z, Myung N (2005) Electrochemistry and electrogenerated chemiluminescence of semiconductor nanocrystals in solutions and in films In: Semiconductor nanocrystals and silicate nanoparticles. Structure and bonding, vol 181. Springer, Berlin, pp 1–57. doi:10.1007/b137239

Chapter 5
Coupling of ECL with Different Techniques

Abstract The perspectives and recent developments in the field of electrochemiluminescence (ECL) have grown exponentially in the last few decades. The state of the art of the developments, key strategies, and trends toward ECL detection coupled to capillary electrophoresis (CE), flow injection analysis (FIA), and solid phase microextraction (SPME) is described. New advances in homemade configurations, designs of ECL flow cells and probes, peculiarities of the operation of tandem systems using miniaturization (microchip/µTAS) with detection by ECL are included. Due to its simplicity, low cost and high sensitivity and selectivity, ECL-based detection has become a quite useful detection tool in CE, FI, and SPME systems, making this technique an interesting field of research. Some emerging phenomena of ECL are discussed, such as light-emitting electrochemical swimmers.

Keywords Capillary electrophoresis · Micellar electrokinetic chromatography · Microchip · µTAS · Nonaqueous capillary electrophoresis · Flow injection analysis · Solid-phase microextraction · Bipolar electrochemical swimmers

5.1 ECL Coupled with Capillary Electrophoresis/Microchip/µTAS

A large number of procedures offering high sensitivity and selectivity are being introduced as the real world applications of electrochemiluminescence (ECL), especially $Ru(bpy)_3^{2+}$-based ECL reactions are continuously gaining considerable interest. Certain features of ECL, namely the necessity of improving the sensitivity, selectivity, and rapidity, and the reduction in costs made it necessary to explore the new ECL instruments integrated with capillary electrophoresis having great potential. Until last few years, capillary electrophoresis has been coupled mainly with fluorescence (FL), electrochemical (EC), spectrophotometric (SP),

S. Parveen et al., *Electrogenerated Chemiluminescence*,
SpringerBriefs in Molecular Science, DOI: 10.1007/978-3-642-39555-0_5,
© The Author(s) 2013

and chemiluminescence (CL) methods; however, since mid-2000s, many researchers are attempting to focus the development of procedures and instruments for capillary electrophoresis with ECL detection.

This chapter spotlights the instrumentation required to generate CE–ECL and new developments of efficient systems for producing ECL coupled with CE which are being used for the growth of highly sensitive ECL assays. Studies focusing some important factors for designing and improving the efficiency and sensitivity of the CE–ECL instrument are also reviewed along with a discussion on the instrumentation of CE–ECL-based assays and related issues. This provides the reader a general understanding of coupling of ECL technique to the CE and its application to biological molecule detection [1]. CE–ECL has taken extensive consideration for last decade after its first discovery and has been widely spread all over the world. A brief discussion on the use of the ECL detectors coupled with different techniques and on the use of CE with the ECL, EC, laser-induced fluorescence (LIF), SP, and CL detectors is also presented in this chapter.

5.1.1 Geographic Sites of Research Centers

ECL in combination with CE is becoming more famous in Eastern Asia, mainly in China and the research centers for the investigation and application of CE–ECL are growing and developing continuously. The first and foremost scientific group is lead by Prof. Erkang Wang in the laboratory of electroanalytical chemistry, Changchun Institute of Applied Chemistry, Chinese Academy of Sciences. Various microfluid ECL instruments developed and applied in France, Canada, and Britain have many significant applications in different areas. The main research centers are listed in Table 5.1. Karkov National University of Radioelectronics (Laboratory of Analytical Optochemometrics), Ukraine also has the setup for CE–ECL investigations. Research is conducted in collaboration with the Kharkov Institute of Physics and Technology (National Science Center); Institute of Solid-State Physics, Material Science, and Technologies (Ukraine National Academy of Sciences); and Institute of Monocrystals supported by the state and is supported by the international projects (UNTTs projects nos. GE 77, 4180, 4495) [1].

In general, ECL has long been employed as a sensitive detection method for flow injection analysis (FIA) and high-performance liquid chromatography (HPLC). The term electrogenerated chemiluminescence is well renowned to emerge in 1929 [2], while articles on the use of ECL phenomena in chromatographic detectors were first appeared in 1980s. ECL combined with FIA systems came on front in 1990s [3, 4] while integrated with the devices for capillary electrophoresis (CE), in 2000. It is worth-noting that direct coupling of ECL to CE is difficult unlike with FIA or HPLC, taking into account that the injection volume in CE is in nanoliters compared with microliters for FIA or HPLC. Specially designed ECL detector is usually required for CE [5]. While coupling ECL with CE, some factors should be kept in mind to improve the sensitivity of the instrument.

Table 5.1 Research centers developing ECL methods in system-on-a-chip design

Research centers	Country
Dalian University of Technology: Laboratory of Precision and NonTraditional Machining Technology. Key Laboratory for Micro/Nano Technology and System of Liaoning Province	China
Chinese Academy of Sciences State (Changchun): Key Laboratory of Electroanalytical Chemistry, Changchun Institute of Applied Chemistry	
Korea University (Seoul): Departments of Biomicrosystem Technology. South Korea	South Korea
University of Tsukuba: Graduate School of Pure and Applied Sciences	Japan
Universite Claude Bernard (Lyon): Institut des Nanotechnologies de Lyon (INL), CNRS	France
Institut de Chimie et Biochimie Moleculaires et Supramoleculaires (ICBMS), Laboratoire de Genie Enzymatique et Biomoleculaire (LGEB),	
Institut des Nanotechnologies de Lyon (INL), CNRS	
University of Science and Technology (Lille), CNRS	
Université-de Montréal Institute for Microstructural Sciences, National Research Council Canada, Département de Chimie	Canada
University of Manchester School of Chemical Engineering and Analytical Science	Great Britain
Institute of Research in the Applied Natural Sciences (Luton)	
University of Neuchâtel: Institute of Microtechnology, Sensors, Actuators and Microsystems Laboratory	Switzerland
Petroleum University of Technology	Iran
University of Athens: Laboratory of Analytical Chemistry, Department of Chemistry	Greece
Cornell University (Ithaca): Department of Biological and Environmental Engineering	United States
Innovative Biotechnologies International, Inc., Grand Island	

1. A source to keep the ECL reagent in contact with the CE effluent from the tip of electrophoretic capillary near the working electrode is required
2. Though the Ru(bpy)$_3^{2+}$ ECL can be well carried out within aqueous or non-aqueous condition; the effects of organic solvents and surfactants on ECL detection should be carefully noted
3. To eliminate or reduce the effect of electric current in CE procedure, ECL cell should be specially designed as it may affect the ECL detection on the microelectrode, different to HPLC coupling with ECL detection [6–8]. CE is holding exceptionally dominating position over high-performance liquid chromatography (HPLC) due to the relative compactness and simplicity of construction of the equipment, rapidity of the analysis, and high-resolution power (the number of theoretical plates reaches 2,000,000). The last parameter is attained using a specific profile of the velocity front in the electro-osmotic and electrophoretic flows in a capillary tube.

It is noteworthy that regarding the suitability of using ECL detectors in CE, even with modern detectors, the potential of CE is often limited by insufficient sensitivity and selectivity because of the extremely small sample volume (10 nm in a capillary 50 μm in diameter), whereas the objects of analysis are determined mainly by the method of detection. This confirms the appropriateness of the CE–ECL system. A number of reviews have been published to recapitulate the progress and key approaches in CE with ECL detection. Few of which are the following: Wang's group in early 2005 published a comprehensive overview regarding CE–ECL systems focusing on the ECL mechanism, CE mode, instrumental design, and ECL efficiency [9]. He along with workers presented another review article in 2007, describing the recent advances and key strategies in CE and microchip CE with ECL and electrochemical detection (ECD), the study is focused on four main parts: CE–ECD; microchip CE–ECD; CE–ECL; and microchip CE–ECL [10]. Rozhitskii and group members summarized the trends of analytical application of CE–ECL especially focused on the advancements and progress in instrumentation for CE–ECL [1]. Lately, Chen and co-workers presented a comprehensive survey of developments in the field of CE–ECL [11]. They illustrated the mechanism involving both co-reactant and inhibitor based ECL, the possible analytes to be determined by CE–ECL, the construction of CE–ECL apparatus and the applications of CE–ECL. Lara with his group members compiled advances and analytical applications in CL and ECL coupled to CE [12].

The increment in the number of publications reveals that the field is still topical and the analytical potential of these instruments is of current interest. Rozhitskii, two years back, compiled publications by year (a), on the use of the ECL detectors with CE, FIA, and chromatography (CH) published from 1960 to July, 2009, and (b), on the use of CE with the ECL, EC, laser-induced fluorescence (LIF), SP, and CL detectors published from 2000 to July, 2009 and found the increased number of publications of CE with ECL year after year.

It is well noted that the detection sensitivity is directly related to ECL efficiency [9]. The first luminol-based CE–ECL detection offered some fundamental design and principle, some of which are stated as under. First, 10 µm carbon or platinum fiber microelectrode aligned with the end of capillary to improve the ECL efficiency and to reduce baseline noise. Second, small inner diameter capillary (<25 µm) was used to restrain the effect from the high electric field on ECL detection. Two optic fibers collect light emitted from the luminol–ECL reaction on the microelectrode which is monitored with a PMT [13]. The instrumentation of CE–ECL is usually exceptionally simple as it lacks the need of external light source. However, coupling CL methods to CE is not an effortless task. The small sample volumes needed in CE separations complicate the addition of CL reagent to the capillary. Therefore, to avoid the complications facing the addition of the CL reagent to the capillary, various strategies have been tried in order to achieve high sensitivity [9].

5.1.2 Addition of the Ru(bpy)$_3$$^{2+}$ Solution

As described earlier, most CE–ECL systems use an end-column Ru(bpy)$_3$$^{2+}$ solution reservoir to supply the ECL reagent for the reason that the flow rate in CE is much lower than that in HPLC. Using an end-column Ru(bpy)$_3$$^{2+}$ solution reservoir, Ru(bpy)$_3$$^{3+}$ was in situ generated at the electrode surface. Thus, reproducibility of the ECL response is affected applying this design which is the main limitation, causing the evaporation of solution over time and the dilution of the solution by CE effluent [9]. Ru(bpy)$_3$$^{2+}$ was added to the CE running buffer to convert to Ru(bpy)$_3$$^{3+}$ at the end of separation capillary. The addition of such a large electrolyte showed some drawbacks of affecting the character of electro-osmotic flow by altering the charge on the capillary wall. Furthermore, it becomes difficult to optimize the detection and separation conditions independently because of the presence of Ru(bpy)$_3$$^{2+}$ in CE running buffer. It is worthy to note that the concentration of Ru(bpy)$_3$$^{2+}$ affects the background signal, linearity, and dynamic range notably [14].

Moreover, the adsorption of Ru(bpy)$_3$$^{2+}$ onto the inner wall of separation capillary increases the equilibrium time to several hours. Figure 5.1 shows a simple apparatus for post-column Ru(bpy)$_3$$^{2+}$ ECL detection, which employed a conductive joint to isolate the separation field from the potential needed to drive the ECL [6].

To prevail over the shortcomings of the above systems, the Ru(bpy)$_3$$^{2+}$ solution can be carried to the cell continuously by a pump for the generation of Ru(bpy)$_3$$^{3+}$ and ECL is attained at the surface of the working electrode. Plexiglass reservoir present at the capillary outlet has an optical fiber that serves for transferring the ECL emission to the PMT. Plexiglass reservoir itself, at the capillary outlet functions as both, the reaction and detection cell for the ECL reaction. Noninterrupted sampling conditions, involving large numbers of samples to obtain good sensitivity and separation efficiency was obtained by optimizing the relative positions of the capillary outlet, working electrode and optical fiber as well as reagent renewal flow-rate. However, in CE–ECL systems, the addition of Ru(bpy)$_3$$^{2+}$ to capillary end using a pump and a mixing tee was less used than in

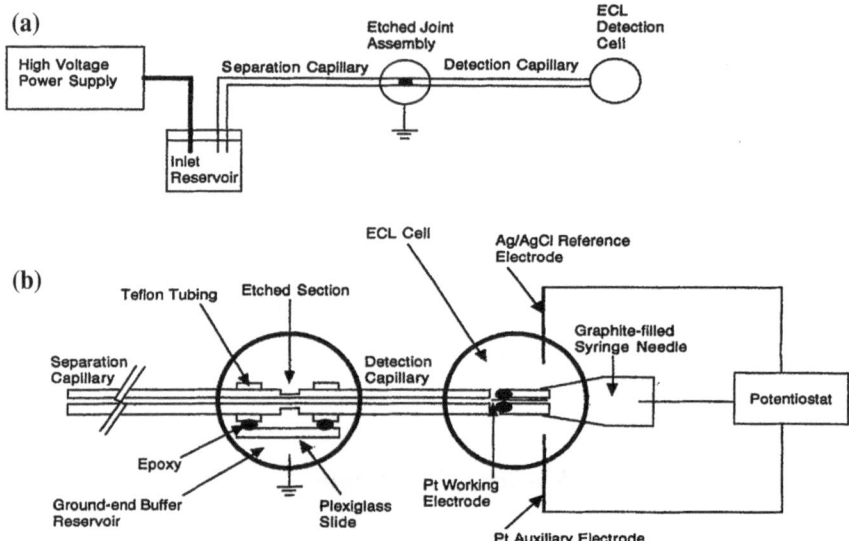

Fig. 5.1 Schematic for capillary electrophoresis with post-capillary ECL detection: **a** Overview of the apparatus and **b** Enlarged view of the etched joint and the ECL detection cell. The PMT (not shown) was positioned directly over the Pt working electrode, and the entire apparatus was place in a fight-tight housing. The ground-end buffer reservoir was filled with 1 0 mM Na_2HPO_4 @H = 9) and the ECL cell was filled with 10 mM NazHP04 (pH = 9) and 1 mh4 $Ru(bpy)_3^{2+}$. Details concerning the fabrication of the etched joint and the ECL cell are provided in the experimental section. Reprinted with permission from Ref. [6]. Copyright 2005 Wiley

HPLC and FIA. This design needs careful optimization of $Ru(bpy)_3^{2+}$ flow rate and the consumption of expensive $Ru(bpy)_3^{2+}$ [7, 15]. Another literature reported a miniaturized chip-type ECL detection cell made-up of two pieces of glass via wet chemical etching. This fabrication requires very low dead volume of the end-column detector of CE and the dilution of the sheath flow to the effluent of CE may reduce the detection sensitivity because of the low flow rate in CE. This design (Fig. 5.2) offers the decreased (to only the volume of the etched channel) dead volume. Instrumental setup is simplified by combining a gravity-induced reagent introduction to an even reagent introduction with flow rate as low as $\sim \mu L\ min^{-1}$ without the use of a pump system [16].

Arora et al. [17] developed a wireless ECL detector to low down the effect of electric current in CE procedure to the ECL detection on the microelectrode, which used the electric field from the separation channel during electrophoretic separation as the required potential difference for the ECL reaction. As shown in Fig. 5.3, a microfabricated "U" shape floating platinum electrode placed across the separation channel, whose two legs functioned as working electrode and counter electrode of the ECL procedure. The device was applied for direct separation and determination of $Ru(bpy)_3^{2+}$ and $Ru(phen)_3^{2+}$ using micellar electrokinetic chromatography (MEKC) on a microfabricated glass device was performed employing this device.

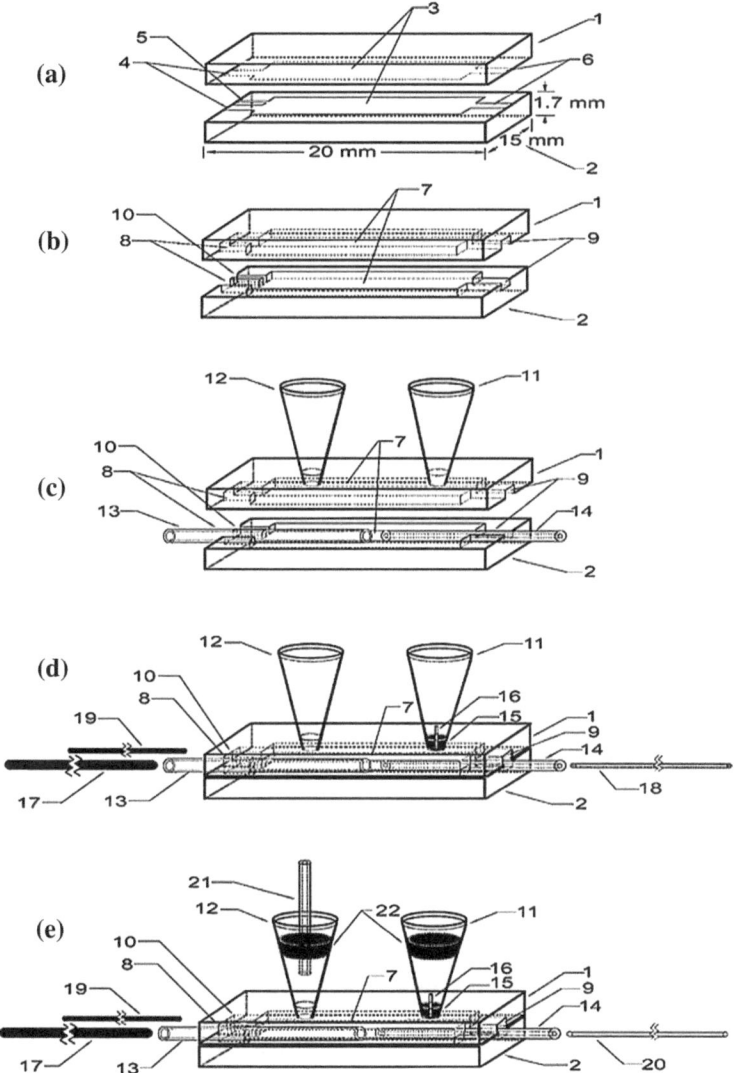

Fig. 5.2 Constructing process of the chip-type ECL flow cell and schematic view of CE and FI measurement modes. Design pattern of flow channel and accommodating grooves (**a**), etched glass plates (**b**), accommodating pipet tips and guides (**c**), CE (**d**), and FI (**e**) measurement modes: cover plate *1*, bottom plate *2*, pattern designs of flow channel *3*, working electrode *4*, reference electrode *5* and capillary *6* accommodating grooves, flow channel *7*, accommodating grooves of working electrode *8*, capillary *9* and reference electrode *10*, Ru-(bpy)$_3^{2+}$ solution reservoir *11*, waste reservoir *12*, working electrode guide *13*, capillary guide *14*, rubber septum [*15*; accommodated at the bottom of the *10*], connecting capillary *16*, working electrode *17*, separation capillary *18*, silver wire quasi-reference electrode *19*, inlet capillary *20*, outlet tubing *21*, and rubber septa seal *22*. Reprinted with permission from Ref. [16]. Copyright 2003 American Chemical Society

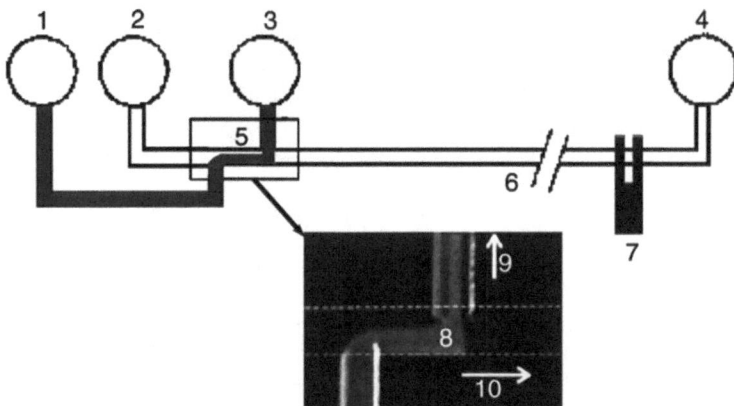

Fig. 5.3 A wireless ECL detection for chip-based CE. *1* sample reservoir; *2* buffer reservoir; *3* sample waste; *4* buffer waste; *5* double-T injector; *6* separation channel; *7* floating platinum electrode (leg width 50 µm, distance between legs 50 µm); *8* sample filled into the double-T injector; *9* vacuum; *10* direction of plug during separation. Adapted from Ref. [17]. Copyright 2001 American Chemical Society

In order to overcome the high field interference, ECL detection was often carried out in an off-column mode to isolate the detector from the CE field, where a field decoupler was created near the capillary outlet. However, the procedures of electrode alignment and decoupler fabrication are always tedious and time-consuming. Here we shall discuss some improvements in the cell design of CE-ECL with and without decoupler [5].

5.1.3 ECL Cell Design

5.1.3.1 ECL Cell Without Decoupler

It is worthy to note that the formation of electronically excited states of electro-chemically generated species takes place on microelectrode which is affected by the electric current present in CE capillaries. Therefore, the effect of CE high electric field on ECL procedure should be abolished. The simplest ECL cell may be the cell without decoupler, but the electric field decoupler has also been used to isolate ECL detections from the CE high voltage field. Some workers reported that using capillary with small inner diameter (≤ 25 µm) or low conductivity buffer, no significant effect was observed from the high electric field on ECL detection. Some methods for the detection of several pharmaceuticals such as diphenhydramine, tramadol, lidocaine, procyclidine, benzhexol, and polyamines have been developed with small inner diameter capillary CE–ECL system. Furthermore in order to obtain high ECL efficiency and to carry out in situ oxidation of Ru(bpy)$_3$$^{2+}$ to

Ru(bpy)$_3$$^{3+}$, the relative position between the working electrode and the end of capillary must be carefully adjusted. Some chemists aligned the microelectrode with the capillary tip, while the other groups inserted the microelectrode directly into the capillary end [9]. For the sake of simplicity of CE–ECL system without decoupler, a group of scientists [18] designed a pseudo-chip CE end-column ECL detection system consisting a flow injection (FI) falling-drop sample introduction interface, a separation capillary of 68 mm, an end-column reservoir integrated with working electrode, counter electrode, reference electrode, and optical fiber fabricated together in a Plexiglas chip [18]. FI analyzer is protected from the CE high voltage employing a FI falling-drop sample introduction interface. The combination of this interface and the short separation time from the short capillary resulted in a sample throughput of 50 h^{-1} [15].

A simplistic and low-cost ECL detection system integrated with CE was introduced without the need of on-capillary field decoupler. A 0.5 mL plastic sample vial glued onto a piece of 1.5 cm × 1 cm × 0.6 mm indium/tin oxide (ITO)-coated glass plate was used for the construction of detector. End-column ECL detection was carried out in a walljet configuration. ECL based on Ru(bpy)$_3$$^{3+}$ reaction was used for the sensitive determination of some trialkyl-amines and amino acids including trimethylamine, triethylamine, tripropylamine, tributylamine, proline, and hydroxyproline. With the purpose of avoiding the possible damage to the sophisticated potentiostat caused due to electrophoretic current flowing into the detector, a simple DC power supply was provided to the ITO electrode (connected in series) and a home-made voltage divider [19].

Afterward, another end-column ECL detector (Fig. 5.4) was developed which lacks the electric field decoupler in CE. This detector was equipped with an electrically heated carbon paste electrode (CPE) where temperature of the electrode (Te) could be controlled by a 100 kHz alternating current (AC) generated from a function generator. The feasibility and reliability of this system was evaluated by applying this system for detection of triethylamine (TEA) and tri-n-propylamine (TPA). Pre-column addition of Ru(bpy)$_3$$^{2+}$ was done into the separation buffer solution. Heated CPE could improve peak shape, gain good reproducibility (lower detection limits), and wider linearity ranges as compared to the conventional electrode at room temperature. As a result, improved linear ranges and detection limits (S/N = 3) for TEA and TPA were obtained when the Te was 397 °C [20].

In order to reduce the labor work of sample pre-treatment, a rapid and sensitive approach for the detection of reserpine in urine (over the range of 1 × 10^{-6} to 1 × 10^{-4} mol L^{-1} with a correlation coefficient of 0.996) was introduced. This method uses MEKC coupled with Ru(bpy)$_3$$^{2+}$ ECL where the sample was analyzed directly. The proposed method uses a 25 µm i.d. capillary as separation column and the ECL detector lacks an electric field decoupler. Field-amplified injection lessens the effect of ionic strength in the sample, increasing the ECL efficiency [21]. The distance of capillary to electrode was also an important parameter for optimizing detection performance as it determined the characteristics of mass transport toward the electrode and the actual concentration of Ru(bpy)$_3$$^{2+}$ in the

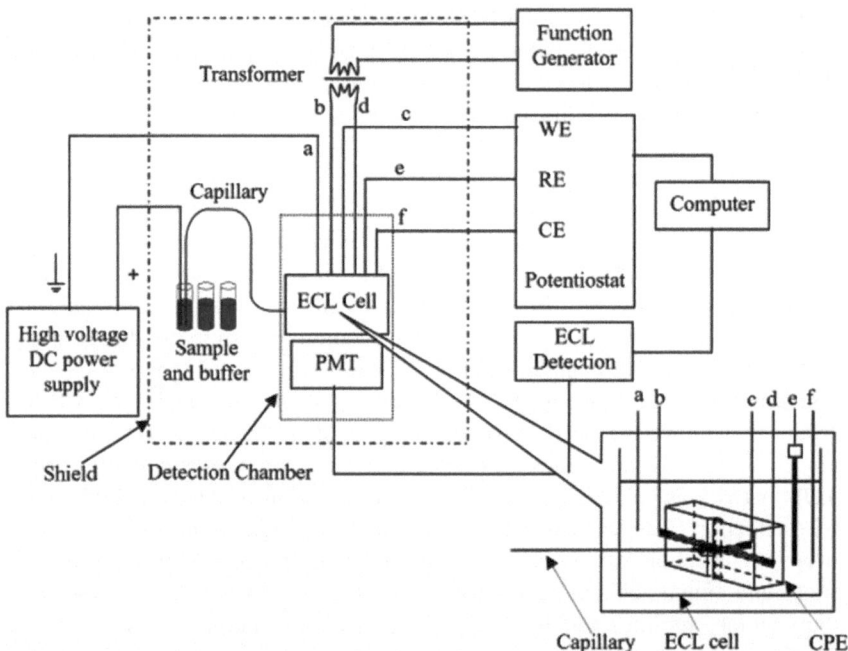

Fig. 5.4 Schematic diagram of the CE–ECL detection system equipped with an electrically heating CPE. *WE* working electrode; *RE* reference electrode; *CE* counter electrode; *a* the grounding of high-voltage power; *b, d* connect to the function generator; *c* connect to the potentiostat; *e* the Ag/AgCl electrode; *f* platinum wire. Reprinted with permission from Ref. [20]. Copyright 2007 Wiley

detection region. A new end-column ECL detection technique coupling to CE was characterized where a 300 μm diameter Pt working electrode was used without an electric field decoupler. This Pt working electrode was directly coupled with a 75 μm inner diameter separation capillary. The hydrodynamic cyclic voltammogram (CV) of $Ru(bpy)_3^{2+}$ displayed no interaction of electrophoretic current toward the ECL reaction. ECL detection potential is shifted due to the presence of high-voltage field. The optimum distance of capillary to electrode is also dependent on the inner diameter of the capillary. For a 75 μm capillary, the distance between the working electrode and the capillary outlet should be within the range of 220–260 μm. A 75 μm capillary in this method, under the optimum conditions offered a linear range for TPA between 1×10^{-10} and 1×10^{-5} mol L^{-1} with correlation coefficient of 0.998 [22].

Ming Li and Sang Hak Lee have reported a CE system integrated with ECL method in the absence of an electric field decoupler for the determination of trimethylamine (TMA) in fish. ECL from the reaction of analyte and in situ generated $Ru(bpy)_3^{3+}$ at electrode surface could be produced in the presence of TMA. For ECL detection, a Pt working electrode biased at 1.23 V (vs. Ag/AgCl) potential was employed in a 10 mmol L^{-1} sodium borate buffer solution, pH 9.2, containing 3 mmol L^{-1} Ru(bpy)$_3^{2+}$. A linear calibration curve (correlation coefficient = 0.9996) was obtained in the range 8×10^{-5} to 4×10^{-8} mol L^{-1} for TMA concentration. Recoveries were acquired in the range 98.78–101.46 %. The system was effectively applied for the assay of TMA in fish, in combination with solid-phase extraction (SPE) disks for sample clean-up and enrichment [23].

To achieve high light capturing efficiency and simplification of the alignment between the separation channel and working electrode, a chip CE was effectively integrated with ECL detection in a recent publication. Working electrode was photo-lithographed by the glass slide coated with indium/tin oxide (ITO). A poly (dimethylsiloxane) (PDMS) layer was reversibly bound to the ITO electrode plate containing injection channel and separation channel. Transparency of ITO electrode helps in achieving the high light capturing efficiency. The experiment unnecessitated the decoupling of ECL detection from the separation voltage using end-column detection mode. Rather the distance between the end of capillary and working electrode should be optimized because the separation electric field decreased the oxidation current of Ru(bpy)$_3^{2+}$ and did not change its oxidation potential significantly. Different to CE–EC detection, where the oxidation current was measured, although the separation electric field had manipulated on the CV, the corresponding ECL intensity did not change significantly within the chip CE [24].

A new CE–ECL detection system operational with an electrically heated Ru(bpy)$_3^{2+}$/multiwall carbon nanotube paste electrode (Ru(bpy)$_3^{2+}$/MWNTPE) was developed (Fig. 5.5). A three electrode $(20 \times 20 \times 20$ mm)-ECL cell was located directly at the top of a PMT and was set in a dark detection chamber. The counter electrode was made of a platinum wire (99.99 %) and the reference electrode of Ag/AgCl in saturated KCl. The Ru(bpy)$_3^{2+}$/MWNTPE was employed as the working electrode, and positioned in parallel direction. The working electrode could be electrically heated and the surface temperature of electrodes could be controlled precisely. These heated electrodes have been shown to provide some advantages over other conventional electrodes at room temperature, such as higher sensitivity, lower RSD, and decreasing width of the peak. Unlike CE–EC detection system where the oxidation current is measured, the corresponding ECL intensity is not affected by the separation electric field in this CE–ECL system. Hence, a wall-jet configuration was adopted, and the end of the capillary was placed directly in the ECL cell without decoupler. Furthermore, wider range of capillary-to-electrode distance and larger-area electrode are a benefit to CE–ECL. The results indicated that the present CE–ECL system coupled with heated modified-electrode could provide high sensitivity, wide linear range, satisfying linear relationship and excellent reproducibility. This system has been successfully applied for the separation and determination of acephate and dimethoate [25].

5.1.3.2 ECL Cell with Decoupler

An instrument with an improved design was introduced later. In this design, the electrophoretic and the electrochemical current were separated by an on-column

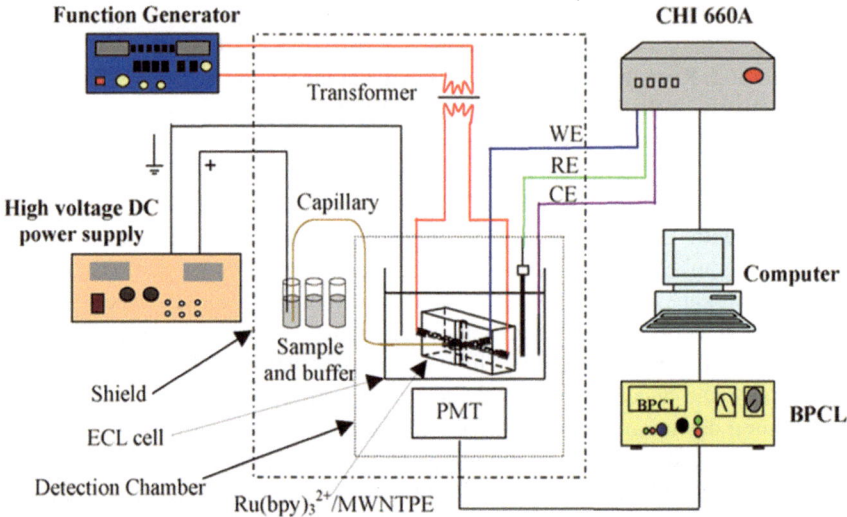

Fig. 5.5 The schematic diagram of the CE–ECL detection system equipped with an electrically heated Ru(bpy)$_3^{2+}$/MWNTPE. *WE* working electrode; *RE* reference electrode; *CE* counter electrode. Reprinted with permission from Ref. [25]. Copyright 2007 Elsevier

fracture which was covered with a Nafion tube. The decoupler used in the above strategy was prepared as follows: ca. 1.5 cm from the end of the capillary was removed after the polyimide coating and the Nafion tubing (1 cm length) was carefully threaded over the score mark. Both ends of the Nafion tubing were sealed to the capillary tubing and a temperate pressure was applied to either end of the Nafion tubing to fracture the capillary at score. Due to wider i.d. of Nafion (0.33 mm) as compared to that of separation capillary (75 or 100 μm), the diffusion of analytes in the fracture resulted in zone broadening and analytes leakage [9].

Wang and Bobbitt investigated the effect of working electrode outer diameter (76 versus 127 μm) on the ECL response and utilized this approach as a detector of MEKC. According to them, the 127 μm o.d. electrode is over 76 μm in terms of LOD, calibration curve slope and correlation coefficient which might be due to the generation of an enhanced amount of Ru(bpy)$_3^{3+}$ in the detection region at larger o.d. electrode. Additionally, larger o.d. electrode facilitates the alignment between the working electrode and the end of capillary [7]. Other literature [14] reported the application of a porous polymer junction near the end of the capillary to complete the CE circuit. EC or ECL current was separated from electrophoretic current by etching the outside wall of the separation capillary. Because of the complexity of another design, a 2 or 5 cm detection capillary was used resulting in peak broadening and decreased separation efficiency. The capillary must be fixed on a plate in order to protect the etched section of capillary. It is noteworthy that the fix of capillary made the fabrication of joint system and the replacement of capillary and working electrode complex; it was also difficult to align the capillary and the working electrode [6].

Accumulated data reveal that two factors result in band broadening in the on-column decoupler, one is the construction of on-column ground section, and the other is length of detection capillary. The on-column etching joint does not create any dead volume [6, 26] as compared to the fracture covered with polymer tubing [7, 8], contributing no band broadening. Because of electric field gradient absence in the detection capillary, sample zone movement is induced from pressure generated in the separation capillary which results in a parabolic flow profile because of the pressure-driven mode [18]. The length of detection capillary contributes drastically to band broadening; hence, a detection capillary as short as possible is favored. Taking the above into consideration, a novel CE–ECL cell different to the previous work [6, 26] was constructed. An instrument where the joint was fabricated by etching the capillary wall with hydrofluoric acid followed by removal of half of circumference of the polyimide coating was developed. The joint here was adequately tough to fabricate the whole system and avoid fixing of the joint on a plate thus reducing the band broadening from the detection capillary [27].

A simplistic, rapid, and efficient analysis of ofloxacin and lidocaine was carried out using a short capillary (10 cm in length) which has the advantage of high sample throughput (60 h^{-1}) and low band broadening in comparison to CE with long capillary. Since no glue was used in this setup to fix, the replacement of capillary and the electrode was much easy; furthermore, the etched joint did not create any dead volume [28]. Another strategy was adapted to couple CE with ECL detection in which the electrical connection of CE was achieved through a porous section at a distance of 7 mm from the CE capillary outlet. A standard three-electrode configuration (Pt wire as a counter electrode, Ag/AgCl as a reference electrode, and a 300 μm diameter Pt disk as a working electrode) was used in the experiment. In comparison with the conventional CE–ECL decoupler designs, this setup with a porous joint has no added dead volume created. Furthermore, it was noticeable that the dead volume can be increasingly decreased by shortening the distance (~ 100 μm) between the working electrode and the end of the separation capillary [27].

Determination of amines, namely triethylamine, tripropylamine, and tributylamine by CE alongwith ECL detection was achieved by a stacking approach based on pH junction and field amplification in order to improve the sensitivity of the system using $Ru(bpy)_3^{2+}$. A two-electrode configuration containing an indium/tin oxide-coated glass as a working electrode and a platinum wire as a pseudo-reference electrode was employed. A flow cell (poly(dimethylsiloxane)- aluminum oxide) made from a mixture of Sylgard 184 silicone elastomer, a curing agent, and aluminum oxide is present in the ECL system. Amines (cations) after preparation in citric acid solution (pH, 4.0) follow the migration toward the background electrolyte (15 mM sodium borate at pH 8.0). This migration gets slow and the amines are stacked at the boundary as a result of deprotonation and causes decrease in the electric field. By applying hydrodynamic injection of the sample for 60 s, this method provides the concentration LODs (S/N = 3) of 24, 20, and 32 nM for TEA, TPA, and TBA, respectively. The results reveal that the stacking CE–ECL system is more efficient than CE–ECL systems using a two-electrode configuration and is comparable to those using a three-electrode configuration. This potential and low-cost CE–ECL system has been applied for the effective determination of 1.0 mM lidocaine, a local anesthetic drug, in urine without any tedious sample preparation [29].

5.1.4 ECL Efficiency

ECL efficiency is another key feature affecting directly the detection sensitivity which is governed by two elements: electrochemistry of $Ru(bpy)_3^{2+}$/analytes at the surface of electrode and related direction between PMT and the electrode where the electrochemistry of $Ru(bpy)_3^{2+}$/analytes performs. Huang et al. illustrated the effect of three different designs among the capillary outlet, working electrode and optical fiber on ECL detection, as illustrated in Fig. 5.6. Figure 5.6b illustrated the sandwich design where the signal was sensitive to the variation of the reagent flow rate. Figure 5.6c shows more sensitivity and separation efficiency; the capillary in this design was located below the working electrode and longitudinally aligned with the optical fiber. In this kind of design, all of the CE effluent may not be in contact with the surface of working electrode [9].

Although the relative positions of capillary outlet, working electrode and PMT were not investigated. Another design, where the optical fiber was replaced with PMT, was used in order to obtain high sensitivity. This difference between Wang's and Huang's work may be from the supply mode of $Ru(bpy)_3^{2+}$. A static $Ru(bpy)_3^{2+}$ reservoir was used instead of Huang's sheath flow mode in this instrument. The dispersion of the ECL emission in all directions and only a part of emission detected was the main drawback of this design [9]. With this design, a high electrochemical efficiency was obtained (a nearly 100 % of the ECL emission generated could be detected by PMT) due to the low dead volume between working electrode and detection window. A Pt wire of 100 μm diameter was wound around the outer wall of separation capillary. After coating the wound Pt wire with epoxy resin to insulate, the Pt ring electrode was obtained by polishing the outlet surface. The integrated capillary/electrode was located opposite to PMT through a transparent window [30].

Forbes et al. [14] introduced working electrode into the separation capillary to perform the ECL and ECL signal was measured using a parabolic reflector to

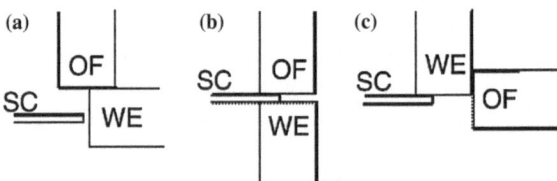

Fig. 5.6 Schematic diagrams of the ECL reaction and light collection system. *SC* separation capillary; *OF* optical fiber; *WE* working electrode. Reprinted with permission from Ref. [9]. Copyright 2004 Elsevier

direct emitted light to PMT. Another sensitive design was also reported where an indium/tin oxide (ITO)-coated glass positioned at the capillary outlet was used as the working electrode. The transparency of ITO glass made all ECL emission detected by PMT. Discrete wavelets transform was used to remove the noise in CE–ECL electropherogram. A new advancement using CE in conjunction with sequential light-emitting diode-induced fluorescence (LEDIF) and ECL detections was presented, after few years, for the determination of alkaloids and amino acids (AAs). In the low cost CE–LEDIF–ECL system, the ECL detector was located in the outlet of the capillary, while the LEDIF detector was positioned 12 cm from the outlet. Naphthalene-2,3-dicarboxaldehyde (NDA) was used to form fluorescent AA–NDA derivatives from AAs possessing primary amino groups, while $Ru(bpy)_3^{2+}$ was used to obtain ECL signals for analytes having secondary and tertiary amino groups. In the presence of poly(ethylene oxide), the CE–LEDIF–ECL separation of a mixture of 12 AA–NDA derivatives, anabasine, nicotine, and proline within 11 min was accomplished. This system allows the analysis of these AA–NDA derivatives and alkaloids at concentrations in the ranges of 49 nM–0.2 mM and 0.66–4.7 mM, respectively. This CE–LEDIF–ECL system offers the advantages of high efficiency, speed, and sensitivity for the analysis of analytes possessing amino groups [31].

CE with ECL detection together with UV spectroscopic and EC methods was used to investigate the chemical oxidation of p-hydroxyphenylpyruvic acid (pHPP) by dissolved oxygen in aqueous solution with high sensitivity and separation efficiency. pHPP was found to be readily oxidized by dissolved oxygen in alkaline solution and yielded a compound strongly enhancing the ECL of $Ru(bpy)_3^{2+}$. The system consists of a detection cell, a Pt ring working electrode, an Ag/AgCl reference electrode, a Pt auxiliary electrode, and a Pt cathode. The detection cell body consists of two pieces of Plexiglas block tightly fixed to each other with a screw. The smaller Plexiglas block (~ 0.5 cm thick) was used only as a transparent window while the bigger one was the main part of the detection cell, which was fabricated differently. This ECL detection system for CE has two apparent advantages: (1) The surface of the working electrode is positioned opposite to the transparent window to detect 100 % of the ECL emission by PMT. It prevails over the weakness of ECL detectors for CE reported earlier, in which the ECL emission scattered in all directions but could only be detected in a very limited angle. (2) The working electrode can be located close to the transparent window to achieve a very small dead volume. The volume of the solution thin layer between the surface of working electrode and the Plexiglas window (shown in Fig. 5.7) is estimated to be only a few nanoliters [30].

It is worth mentioning that a europium(III)-doped prussian blue analog (Eu-PB) film was modified chemically on the surface of a microdisk platinum working electrode to avoid the possible electrode fouling as well as to improve the ECL efficiency and detection sensitivity. After optimizing the conditions, the ECL intensity was in proportion to analyte concentration in the range from 0.01 to 1.0 μg mL^{-1} with a detection limit of 2.0 ng mL^{-1} (3σ). This CE–ECL method based on $Ru(bpy)_3^{2+}$ system had been applied for the determination of sinomenine

Fig. 5.7 Schematic diagram
of electrochemiluminescence
detection for capillary
electrophoresis. The working
electrode arrangement.
Reprinted with permission
from Ref. [30]. Copyright
2003 American Chemical
Society

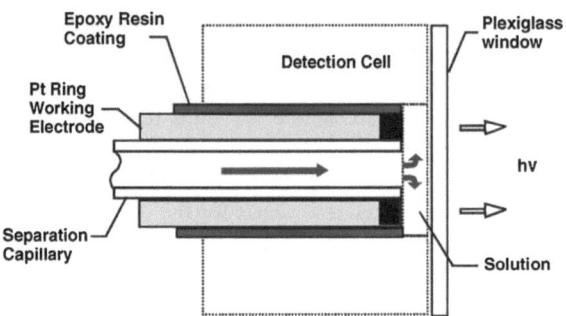

in Chinese herb S. acutum Rehd. et Wils [32]. This method is found to be
promising not only as a good alternative for the rapid determination of sinomenine
in plant extracts with good selectivity, wide linearity, and reliable stability but also
found a proficient ancillary technique for the preliminary exploration of other
quinolizidine alkaloids in Chinese traditional herbs.

In the following year, analysis of Ascorbic acid (AA) in single rat hepatocyte
cells by coupling CE with ECL was reported, where a homemade CE–ECL device
is employed for the determination of a single cell at a carbon fiber microdisk
bundle electrode with high sensitivity of ECL as shown in Fig. 5.8. $Ru(bpy)_3^{2+}$
(4.0×10^{-3} M) was injected into the capillary tip after a cell lysed and therefore
AA in individual rat hepatocytes reacted with $Ru(bpy)_3^{2+}$ at the working electrode.
After the application of a separation voltage of 15 kV across the capillary, ECL
response was recorded. The present system offers the advantage of individual rat
hepatocyte cell analysis directly without any pre-treatment. Moreover, owing to
the small sampling and fast separation capabilities of CE and high ECL sensitivity,
the developed method is useful and powerful for chemical species analysis in
single cells [33].

Procyclidine is a tertiary amine compound which has been used to treat
Parkinson's disease. Several analytical methods namely capillary gas chromatog-
raphy, gas chromatography, and FIA have been used for the assay of procyclidine.
Another more sensitive and selective method illustrating a $Ru(bpy)_3^{2+}$-based
CE–ECL approach has been presented to detect procyclidine in human urine.
An ECL detection cell for post-column addition of $Ru(bpy)_3^{2+}$ was designed. Under
optimized conditions, a detection limit of 1×10^{-9} mol L^{-1} in an on-capillary
stacking mode is achieved. For application in urine, a cartridge packed with slightly
acidic cation-exchange resin was used to eliminate the matrix effects of urine and
improve the detection sensitivity with an extraction recovery of 90 % [34].

Simultaneous EC and ECL detection of amphetamines namely methamphet-
amine, 3,4-methylenedioxyamphetamine, and 3,4-methylenedioxymethamphet-
amine was performed employing CE. Factors affecting the separation and
detection performance, such as the detection potential, the pH value and concentration of the running buffer, the separation voltage, and the pH of the detection
buffer, were explored. A liquid–liquid extraction with ethyl acetate procedure was

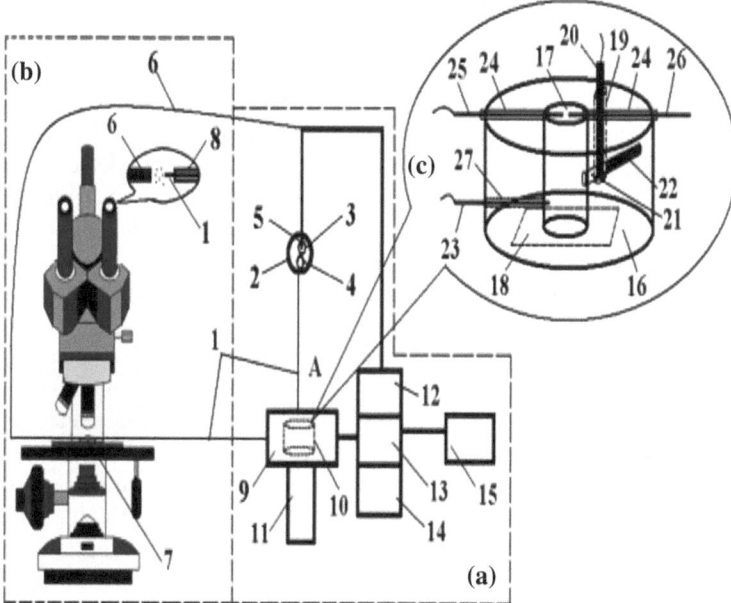

Fig. 5.8 Overview of the CE–ECL detection and the single-cell analysis system. *A* CE–ECL detection system; *B* single-cell injection device; *C* the enlargement of the CE–ECL detection cell. *1* Capillary; *2* support of sample reservoir and running buffer reservoir; *3* running buffer reservoir; *4* sample reservoir; *5* high-voltage anode; *6* copper wire; *7* setup of single cell injection [15]; *8* steel tube; *9* light-tight box; *10* CE–ECL detection cell; *11* PMT; *12* high voltage; *13* electrochemical potentiostat; *14* ECL data collector; *15* computer; *16* poly- (methyl methacrylate) stick; *17* buffer reservoir; *18* microscope slide; *19* reference electrode port; *20* reference electrode; *21* connecting passage between (*17*) and (*19*); *22* epoxy resin; *23* auxiliary electrode and high-voltage cathode; *24* steel tube; *25* working electrode; *26* capillary; *27* auxiliary electrode passage. Reprinted with permission from Ref. [33]. Copyright 2008 Wiley

introduced for urine sample pre-treatment for practical application, and obtainable extraction efficiencies were higher than 90 % [35]. For the estimation of chloroquine phosphate (CQ), a CE method with endcolumn ECL detection has been reported. Parameters affecting detection were optimized, leading to a detection limit of 3×10^{-7} M (S/N = 3), and a linear calibration curve was obtained for the system over two orders of magnitude. Chloroquine phosphate, difenidol hydrochloride, and clomifene citrate were successfully separated at pH 7.0 [36]. The determination of gatifloxacin in biological fluid is performed by a simple, rapid and sensitive CE method based on the ECL reaction of gatifloxacin and Ru(bpy)$_3^{2+}$ in the end-column detection cell. The proposed method scheme has been effectively applied to determine gatifloxacin in urine and blood samples after clean-up process using C18 SPE column. As compared to nonchromatographic methods, the proposed approach is fit for biological samples as it possesses satisfying selectivity and avoids interference brought by the complex sample matrix for the reason of the effective clean-up and separation process. The proposed method is promising and offers wider range of practical applications in pharmacokinetics and pharmacology investigations [37]. It is worthy to note that clindamycin-enhanced ECL intensity of $Ru(bpy)_3^{2+}$ as a coreactant. This finding led to the development of CE coupled with end-column ECL for the quantitative detection of clindamycin. After optimization of the experimental conditions, the ECL intensity showed linearity in the concentration range of 5.0×10^{-7} to 1.0×10^{-4} M with a detection limit of 1.4×10^{-7} M for clindamycin. The application of this CE–ECL technique was found for the detection of clindamycin in pharmaceutical and clinic samples. Studies on the interaction of clindamycin with hemoglobin were also explored where the binding constant of clindamycin with hemoglobin was estimated to be 3.6×10^3 M^{-1} [38].

Galanthamine (GAL) in Bulbus Lycoridis Radiatae has also been determined by coupling CE with an end-column $Ru(bpy)_3^{2+}$ ECL. Under the optimized conditions, this determination can be performed within 6 min with lower detection limit and wider linear range. It is suggested that CE–ECL is suitable for study of the components of plant extracts and can possibly become a substitute or supplement to HPLC-based methods. This method has potential to be used in pharmaceutical industry for the quality control of raw materials [39]. Furthermore, pseudolycorine in the bulb of lycoris radiate has been determined by CE coupled with online ECL detection and ultrasonic-assisted extraction. The method offers fast separation, low LOD, good selectivity, and high sensitivity, and the proposed method is suitable and accurate for the detection of Amaryllidaceae alkaloids in plant extract [40]. For the determination of josamycin (JOS), a novel and sensitive method based on CE–ECL detection has been described where platinum disk electrode (300 μm in diameter) was used as a working electrode. Josamycin in rat plasma can be determined in 6 min by this method, and the extraction recoveries with spiked plasma samples were over 92 %. JOS is determined based on the high-enhancement effect on the $Ru(bpy)_3^{2+}$ due to the presence of tertiary amine group, since tertiary amine has the feature of ECL enhancement. The method can also be applied to other compounds detectable by $Ru(bpy)_3^{2+}$ in both biological and nonbiological samples [41].

Recently, CE coupled with $Ru(bpy)_3^{2+}$ ECL for highly sensitive detection of metformin hydrochloride (MH) derivatized with acetaldehyde is reported. The precolumn derivatization of MH with acetaldehyde was performed in phosphate buffer solution (0.3 mol L^{-1}, pH 7.5) at room temperature for 120 min. The factors affecting this method of analysis, e.g., acetaldehyde concentration, buffer pH, electrokinetic voltage and injection time were examined and under optimization of these conditions, the MH ECL detection sensitivity was more than 120 times that without derivatization. The detection of 0.3 ng mL^{-1} with S/N = 3 was attained. The proposed method is simple, economical, and sensitive and is used for the determination of MH in urine [42]. A method combining CE with $Ru(bpy)_3^{2+}$ ECL detection that can be applied to amine-containing clinical species was developed, and the performance of CE–ECL as a quantitative method for the determination of sulpiride in human plasma or urine was evaluated [43].

5.1.5   Microchip Capillary Electrophoresis/μTAS

Various separation techniques, for instance, capillary electrophoresis (CE), microchip electrophoresis (ME), high-performance liquid chromatography (HPLC), and laser-induced fluorescence (LIF) have been integrated with $Ru(bpy)_3^{2+}$ ECL. During the past several years, $Ru(bpy)_3^{2+}$ ECL detection coupled with CE and ME separation is acquiring dominant position and has been proven to be more efficient and attractive system for determination of complex samples. In comparison with laser-induced fluorescence or mass spectrum detection which are commonly employed on CE and ME, ECL detection presents comparable sensitivity, high flexibility and relatively low cost. However, there are not many publications on the microchip CE combined with ECL detection. In 2007, Wang's group presented a review to recapitulate the advances and key strategies in CE and ME with EC and/or ECL detection [44].

In the preliminary studies, a number of microchip CE devices were constructed from expensive glass and quartz due to the established micromachining technology of these materials. This approach remained persistent only for few years and was replaced by alternate materials such as plastics and polymers due to their high cost and complicated fabrication procedures. A broad range of polymer materials, such as polycarbonate [45] and poly(methyl methacrylate), are found to have the property of microchip fabrication. In particular, poly(dimethylsiloxane) (PDMS), as one of these polymer materials, has been extensively studied and reviewed by Whitesides and co-workers [46, 47]. Effenhauser [48] gave first report on the microchip CE device that combined a PDMS layer and a glass substrate. PDMS has been broadly used and well-studied owing to a number of advantages, such as low cost, rapid prototyping complex device and its reversible/irreversible binding to smooth surfaces. In addition, the fabrication procedures based on rapid prototyping and replica molding are accessible to a general laboratory without ultraclean conditions [49]. Likewise, two-channel and multichannel microfluidic sensors based on this principle were also described in their later work [50, 51].

The field of micro total analysis systems (μTAS), also called "lab-on-a-chip", has also been growing rapidly [52, 53]. Considerable interest has been focused on μTAS and particular attention has been paid to CE microchips. Most of the detection methods employed in conventional CE are adopted in the microchip layout. Owing to several advantages, such as high sensitivity, simplicity, low cost, and ease of miniaturization make ECL powerful detectors for microchip CE. Some other reviews partly deal with the detection methods coupled to CE and microchip CE [10]. Manz's group first illustrated an integrated microfluidic system with a U-shaped floating platinum electrode ECL detector. The two legs of U-shaped floating platinum electrode functioned as working and counter electrode and the platinum electrode was placed across the separation channel. The detection potential for the ECL reaction is generated directly from the electric field for electrophoretic separation [17]. Crooks and his co-workers reported a novel microfluidics-based sensing system that based on EC detection and ECL reporting.

In their work, the ECL reporting reaction was chemically decoupled from the electrochemical sensing reaction [54]. In the same year, a novel reagentless solid-state CE detector was fabricated by immobilizing $Ru(bpy)_3^{2+}$ in a polymer film coated onto a Pt disk electrode [55]. In the following year, Fang's group constructed a new miniaturized CE–ECL post-column detection system by coupling with FI sample delivery system, as shown in Fig. 5.9. They used a falling-drop interface to carry out FI split-flow sample introduction [15].

A new ECL chip-type detection cell, which could be used for both CE and FIA was constructed by Wang and co-workers [16]. In their design, the chip-type detection cell is easily convertible to CE and FI measurement mode with run-to-run RSDs of less than 2.5 % for both CE and FI measurements in the linear concentration ranges. This design made a contribution in the development of ECL assay for clinical and biological applications. Subsequently, they fabricated another CE–ECL system with a porous joint as electrical connection of CE. This joint was fabricated by etching capillary wall with hydrofluoric acid after the removal of polyimide coat. Because the porous capillary wall allowed the CE current to pass through and there was no electric field gradient beyond that porous section, the influence of CE high electric field on ECL procedure was negligibly reduced, and no added dead volume was created [27]. Same group has reported a series of researches about an integrated ITO electrode-based $Ru(bpy)_3^{2+}$ ECL detector for a PDMS/glass hybrid MCE device. This MCE–ECL strategy utilizes an ITO-coated glass slide as the chip substrate with a photolithographically fabricated ITO electrode located at the end of the separation channel in a PDMS layer [56].

Microfluidic chip based on CE has attracted extensive attention recently. Its advantages included short analysis time, portability, disposability, and minute sample and reagents consumption. Moreover, the integration of sample-handling, separation, and detection makes the chip CE system much simplistic and user-friendly. In 2003, the first report about an integrated indium tin oxide electrode-based $Ru(bpy)_3^{2+}$ ECL detector for a PDMS microchip CE device was presented which describes an indium tin oxide electrode based $Ru(bpy)_3^{2+}$ ECL detector for a microchip CE. The microchip CE–ECL system consists of a PDMS layer which contains separation and injection channels and an electrode plate with an ITO electrode made by employing photolithographic method. Photon-capturing efficiency is greatly enhanced by binding PDMS layer reversibly to the ITO electrode plate, which greatly simplified the alignment of the separation channel with the working electrode. Moreover, high separation electric field has negligible significant influence on the ECL detector, and decouplers for isolating the separation electric field were not needed in the microchip CE–ECL system. This microchip CE–ECL strategy was successfully applied to proline and tripropylamine as model analytes [24].

In the following year, lincomycin was determined by the microchip CE–ECL system where ITO working electrode was fabricated by photolithographic method from an ITO-coated glass slide (chip substrate) located at the end of the separation channel (Fig. 5.10). The top layer made up of a poly(dimethylsiloxane) (PDMS) layer consisting of two channels, namely separation and injection channels. This microchip CE–ECL system can be successfully applied for the rapid analysis of

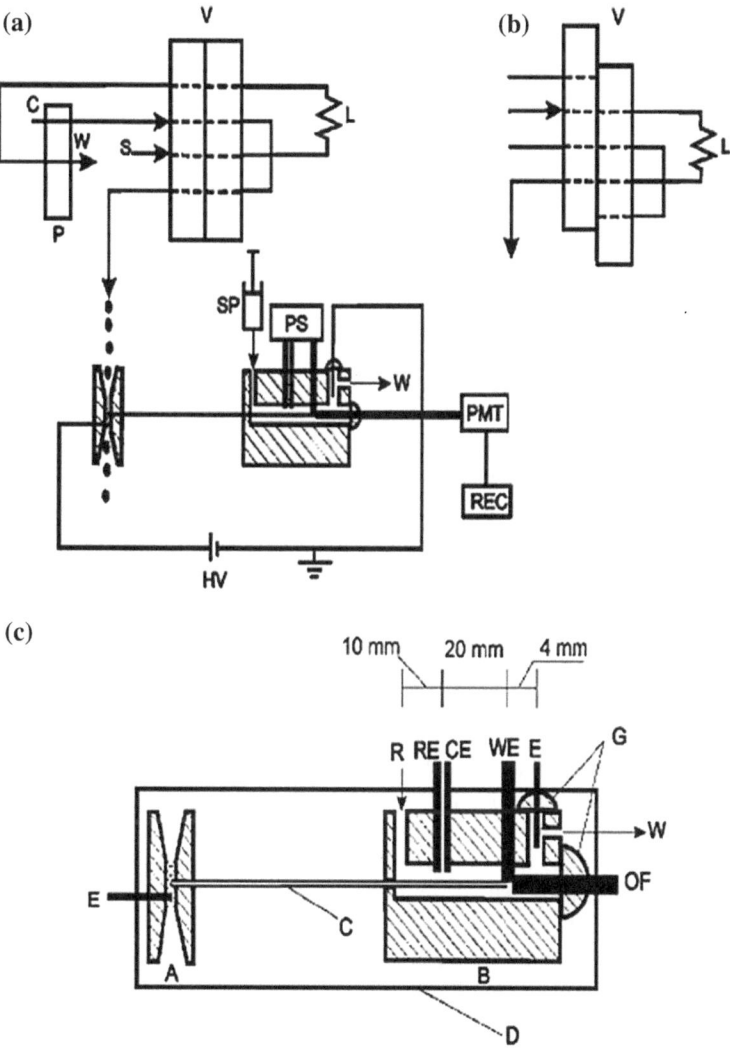

Fig. 5.9 Experimental setup of the miniaturized CE system with FI sample introduction and ECL detection: **a** sample loading; **b** injection. *S* sample; *C* carrier; *W* waste; *P* peristaltic pump; *SP* syringe pump; *V* eight-channel 16-port valve; *L* sample loop; *HV* high voltage supply; *PS* potentiostat. **c** Schematic diagram of the microfluidic device. *A* falling-drop interface (inlet tubular reservoir); *B* ECL reaction and detection cell (outlet reservoir); *C* separation capillary; *D* chip baseplate; *E* Pt wire electrodes; *RE* Ag/AgCl reference electrode; *CE* counter electrode; *WE* Pt working electrode; *G* epoxy glue; *OF* optical fiber; *R* ECL reagent; *W* waste. Dimensions not to scale. Adapted from Ref. [15]. Copyright 2002 Elsevier

lincomycin, a tertiary amine, within 40 s in a urine sample without pre-treatment. Linear range of 5–100 μM and the LOD of 3.1 μM were obtained with correlation coefficient of 0.998, respectively, under the optimized conditions. This strategy

Fig. 5.10 Schematic
diagram of the microchip
integrated with ITO working
electrode. Reprinted with
permission from Ref. [57].
Copyright 2004 Elsevier

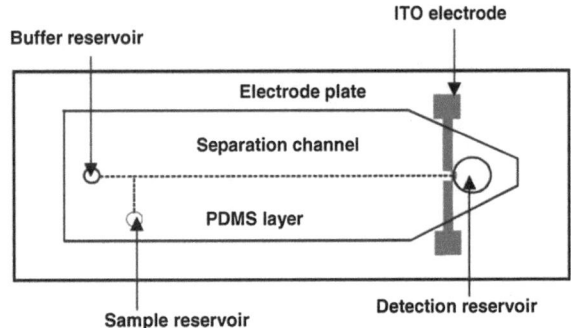

does not need decoupler for isolating the high separation electric field because the
influence of the high separation voltage on the ECL signal was negligible. Owing
to a number of features, such as simplicity, selectivity, sensitivity and low sample
consumption, this strategy has been found to offer unique advantages in biochemical applications and for determination of medicines in clinical analysis [57].

Next year, a tertiary amine derivative, 2-(2-aminoethyl)-1-methylpyrrolidine
(AEMP) was successfully exploited as ECL probe within microfluidic chip using
ECL detection. The system was based on the interaction between biotin and
avidin. Firstly, the ECL efficiency of AEMP was characterized via comparing with
that of two coreactants enhancing $Ru(bpy)_3^{2+}$ ECL, TPA and proline. At the same
condition, AEMP has a similar ECL efficiency to TPA, and much higher than
proline which indicates that AEMP has a good reactivity to the analytes containing
carboxyl group with a similar ECL efficiency to TPA. After reaction of AEMP
with NHS-LC-biotin (succinimidyl-6-(biotinamido) hexanoate), analysis of the
products and their ECL was performed by directly injecting it in the microfluidic
chip and the mixture of AEMP and biotinylated AEMP was separated using a
4.5 cm microchannel. Under optimal condition, the detection limit (S/N = 3) of
AEMP was achieved as 2.7 μM [58].

Furthermore, simultaneous EC and ECL dual detection system to microchip CE
was also introduced by Qiu et al. [59], in which $Ru(bpy)_3^{2+}$ was used as an ECL
reagent as well as a catalyst (in the formation of $Ru(bpy)_3^{3+}$ for EC detection. In
$Ru(bpy)_3^{2+}$ ECL process, $Ru(bpy)_3^{3+}$ generation followed by reaction with analytes resulting in an ECL emission was noticed with a great current enhancement
in EC detection owing to the catalysis of $Ru(bpy)_3^{3+}$. The current response and
ECL signals were examined simultaneously. This dual detection strategy was
validated in case of dopamine, anisodamine, ofloxacin, and lidocaine. Usually, for
the EC detection of dopamine (DA) with the presence of $Ru(bpy)_3^{2+}$, a ~5 times
higher signal/noise ratio can be achieved than that without $Ru(bpy)_3^{2+}$, during the
concurrent EC and ECL detection of a mixture of DA and lidocaine using CE
separation. In this experiment, it was found that a simple and convenient detection
method for analysis of more kinds of analytes in CE separation was achieved
employing this dual EC and ECL detection strategy rather than the single EC or

ECL detection alone, and additional information of analytes could be gathered in analytical applications simultaneously.

One of the several other strategies to reduce the consumption of expensive ECL reagent includes the contribution of Du et al. They constructed a solid-state ECL detector joined with microchip CE by immobilization of $Ru(bpy)_3^{2+}$ into Eastman AQ55D-silica-carbon nanotube composite thin film on a patterned ITO electrode [60]. In the following year, Chen's group fabricated a facile and universal wall-jet configuration approach for the microchip CE–ECL detection system. In this approach, both pre-column and post-column detection modes were made functional to determine TPA, tramadol, and lidocaine [61]. In the same year, same group [62] explained a wall-jet configuration for MCE–ECL detection system, where the working electrode (400 μm-diameter glassy carbon) was aligned to the outlet of separation channel and set by a three-dimensional adjustor under the microscope. In the meantime, the end of separation channel outlet of the glass microchip facilitated the working electrode alignment by rubbing it away to a wedge shape with the thickness of 500 μm. Though, the process of fabricating solid-state ECL sensors is tedious; these sensors were also attempted for application in MCE–ECL. Recently, an efficient and easily fabricated ECL detection system with a simplistic and compact electrode configuration for microchip CE was described using a 300 μm-diameter platinum disk working electrode, entrenched in a titanium tube which acts as counter electrode. The successful functionality of the method was confirmed by separation and determination of proline and tripropylamine which is linear in the range from 5 to 5,000 μM for proline with a detection limit of 1.0 μM (S/N = 3). The system also has applicability in the determination of chlorpromazine hydrochloride in pharmaceutical formulations Cutaway view of the integrated working-counter electrode is shown in Figs. 5.11 and 5.12 displays microchip fabrication, construction of guiding channel and a part of the detection cell, and detection cell assembly [63].

For sensitive and selective quantitative chemical analysis, a miniaturized hot electron-induced ECL device with subnanomolar detection limits was fabricated where both the working and counter electrode are integrated into a single chip. In this study, three schemes were examined for sample containment, consisting of a separately assembled PTFE sample cell; integrated PDMS sample chambers, and hydrophobic sample confinement (Fig. 5.13). All the three sample containment schemes have optimal electrode geometries and the uniformity of ECL in each was investigated. This method offered excellent linearity and subnanomolar detection limits with the model analyte. Maximal working electrode area was required in

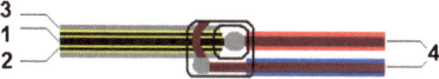

Fig. 5.11 Cutaway view of the integrated working-counter electrode: *1* platinum wire, *2* glass tube, *3* titanium tube, *4* cable. Dimensions are not in scale. Reprinted with permission from Ref. [63]. Copyright 2011 Springer

Fig. 5.12 a Microchip fabrication: *1* sample reservoir, *2* sample waste reservoir, *3* buffer reservoir, *4* separation channel; **b** Construction of guiding channel and a part of the detection cell: *5, 6* a piece of soda lime glass, *7* a part of the detection cell, *8* guide channel for electrode; **c** The whole detection cell assembly: *9* a microscope slide, *10* the whole detection cell. Reprinted with permission from Ref. [63]. Copyright 2011 Springer

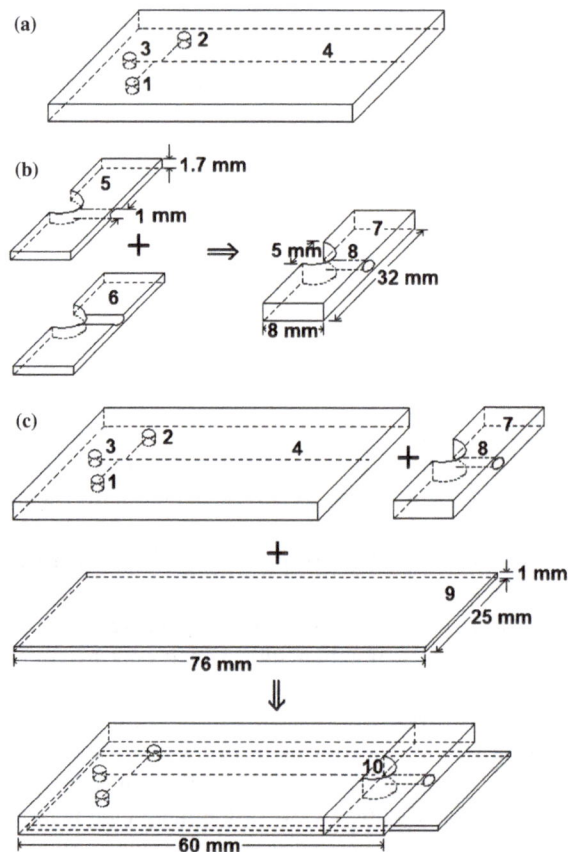

Fig. 5.13 Planar silicon and glass electrode chips and the PTFE sample cell (*left*), a silicon device with a capillary-filling PDMS chamber (*middle*), and a silicon device with a hydrophobically confined sample droplet (*right*). Reprinted with permission from Ref. [64]. Copyright 2010 Elsevier

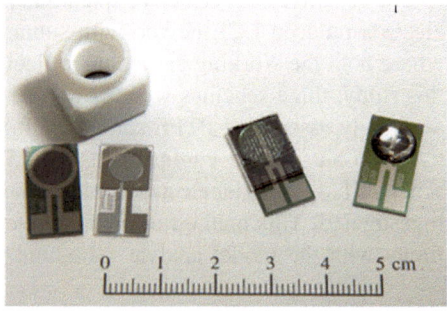

case of luminescence efficiency, whereas in the PDMS and hydrophobic versions a sufficiently dense mesh-type counter electrode was found important to maintain uniform luminescence over the electrode [64]. Figure 5.14 summarizes cross-sectional views of silicon (A) and glass (B) devices.

Fig. 5.14 Cross-sectional
views of silicon (**a**) and glass
(**b**) devices. Vertical and
lateral dimensions are not to
scale. Reprinted with
permission from Ref. [64].
Copyright 2010 Elsevier

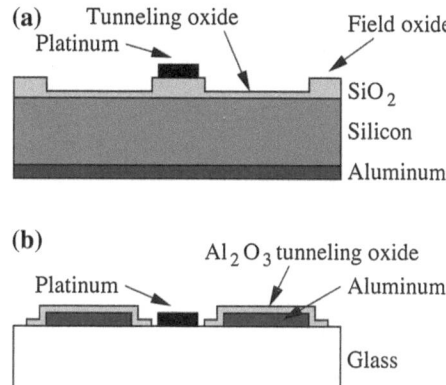

And for rapid analysis of tertiary amines, simultaneous ECL and EC detection scheme coupled with nonaqueous capillary electrophoresis (NACE) was developed employing Pt electrode. This NACE–ECL/EC dual detection strategy was utilized for the analysis of triethylamine, tripropylamine, chlorpromazine, promethazine, and dioxopromethazine (DPZ). The results show that NACE–ECL/EC had the advantages of simple and fast analysis with more information, wide linear range, high sensitivity, and compatibility with real urine sample. NACE–ECL/EC system was thought to be a potential method for clinic sample analysis [65]. Crooks' group describes a two-channel microfluidic sensor that uses anodic ECL as a photonic reporter of cathodic redox. An approach further expanding the number of analytes that can be detected was developed, because even those that might interfere with the ECL cocktail are accessible. Furthermore, this two-channel strategy validates the possibility of using completely different solution conditions (for example, different solvents or electrolytes) in the detection and reporting channels [66].

Benzhexol hydrochloride is an antimuscanic agent and is used for the symptomatic control of all forms of Parkinsonism. Various analytical methods for its determination are present in literature, such as PVC membrane ion-selective electrode, derivative spectroscopy, inductively coupled plasma atomic-emission spectroscopy, gas chromatography, liquid chromatography, and high performance thin layer chromatography. Some of which offered limited sensitivity while other methods required expensive apparatus. Keeping all this in mind, Wang [67] and Yang's [68] group combined capillary zone electrophoresis (CZE) with an ECL [67] and with end-column ECL detection system [68] and validated it to the determination of benzhexol hydrochloride in body fluids with low detection limit (6.7×10^{-9} M (S/N = 3)) and bisoprolol in tablets and drugs [67]. Bisoprolol reacts as a co-reactant in $Ru(bpy)_3^{2+}$ ECL system and tetrahydrofuran is utilized as an additive in the running buffer to obtain the absolute ECL peak of bisoprolol. The proposed CZE–ECL method has the advantage of requiring aqueous solutions and a shorter migration time for bisoprolol in comparison with CZE–UV as well as

HPLC. The proposed method is rapid, low cost, simple, and sensitive and is promising to determine all beta-blockers, especially to monitor bisoprolol in clinical and biochemical samples in future [68].

CEC, a hybrid separation technique of CE and HPLC, potentially combines the high efficiency of CE and the selectivity and versatility of HPLC. CEC was coupled with $Ru(bpy)_3^{2+}$ ECL detection by Xu's group for the first time. Proline, putrescine, spermidine, and spermine were separated efficiently by this method at pH range of 3.5–7.0 of the mobile phase. The proposed detection scheme offers a powerful tool for analysis since CEC generally displays better resolution than CZE for neutral analytes and charged analytes with equal electrophoretic mobilities [69]. Dual-cloud point extraction (dCPE) and tertiary amine labeling has been used for sensitive and selective detection of two auxins employing CE–ECL. The proposed method shows feasibility to the detection of the two auxins in acacia tender leaves, buds, and bean sprout and the detection limits (S/N = 3) were 2.5 and 2.8 nM for indole-3-acetic acid (IAA) and indole-3-butyric acid (IBA), respectively. The proposed method was applied effectively to the detection of the two auxins in real samples [70].

Sensitive determination of various amine-containing analytes including pharmaceuticals, purines, and DNA without any derivation was performed by combining gold nanoparticles with CE $Ru(bpy)_3^{2+}$ ECL [71]. CE–ECL using acetaldehyde as a new derivatization reagent is also used for the analysis of primary and secondary amino acids with selective derivatization. These amino acids were derivatizated pre-column with acetaldehyde which was performed in aqueous solution at room temperature for 1 h. Arginine, proline, valine, and leucine were used as model compounds and their derivatives could be completely separated by CE and sensitively detected by ECL within 22 min. Their detection limits of 1×10^{-7} (0.5 fmol, arginine), 8×10^{-8} (0.4 fmol, proline), 1×10^{-6} (5 fmol, valine), and 1.6×10^{-6} M (8 fmol, leucine) were obtained at S/N = 3. Acetaldehyde, as a new derivatization reagent as compared to other derivatization reagent is simple, commercially available, and could be applicable to a various drugs and compounds containing a secondary or primary amino group [72].

Simultaneous analysis of six cardiovascular drugs (alprenolol, propafenone, acebutolol, verapamil, atenolol, and metoprolol) via central composite design was performed coupling CE with dual, EC and ECL detection. Moreover, three critical electrophoretic factors employed in this method, namely Tris-H_3PO_4 buffer concentration, buffer pH value, and separation voltage were investigated. The developed CE–EC/ECL system was validated in urine sample for rapid, sensitive, accurate, and reproducible analysis [73].

5.2 ECL Coupled with Flow Injection Analysis

A glucose biosensor based on luminol ECL, where sol–gel method is employed to immobilize glucose oxidase (GOD) on the surface of a glassy carbon electrode was illustrated. Carrier, luminol and sample solutions are delivered by a peristaltic pump. The thin flow cell consists of an Ag/AgCl reference electrode, a stainless steel counter electrode, and a polytetrafluoroethylene (PTFE) block fitted with a working electrode and a plexiglass window. ECL light emission is collected and transported by a quartz optical fiber bundle which is attached to the PMT of a luminometer at one end. After optimizing the working conditions, glucose could be detected with a detection limit of around 26 mM. This novel FI optical fiber biosensor showed good selectivity and operational stability[74].

For the detection of glucose, a thin-film glucose ECL biosensor [75] and ECL-simple planar optical waveguide connected to a FIA system (detection limit of 0.3 mM) [76] is reported. In the former approach, thin film technology was adopted in order to achieve faster response and to obtain the concept of miniaturization biosensor. The method was based on $Ru(bpy)_3^{2+}$ doped in alcohol-free low-volume shrinkage mesoporous silica sol–gel with PEG-400 as the template. In the latter one, light generated by luminol ECL is collected with a photomultiplier tube and a photon counter unit. The waveguide (covered with ITO and modified covalently with GOD) is mounted to a thin-layer cell which is connected to a FIA system [76]. The luminol ECL-FI-fiber optic based biosensors were used for glucose and lactate measurements in sera and also for lactate measurements in whey solutions. Figure 5.15 shows schematic representation of the flow cell for the above ECL measurements and detailed representation of the sensor/flow cell interface [77].

Same group in the following year integrated ECL with FIA using GOD or lactate oxidase immobilized on polyamide membranes. H_2O_2 is generated in the presence of oxidase, which causes the light emission to trigger electrochemically by means of a glassy carbon electrode polarized at +425 mV versus a platinum pseudoreference electrode. Glucose and lactate were determined with the detection limits of 150 and 60 pmol, respectively, and the dynamic ranges achieved were linear from 150 pmol to 600 nmol and from 60 pmol to 60 nmol, respectively [78]. Indirect determination of glucose is also reported, integrated with a FIA system. The strategy is based on the linear relationship between concentration of H_2O_2 and the decrease in ECL intensity in a $Ru(bpy)_3^{2+}/TPA$ system. The proposed method has the potential of developing new analytical procedures for the analyses of clinical and biomedical related analytes which produce H_2O_2 as their enzymatic product with the $Ru(bpy)_3^{2+}/TPA$ ECL system [79]. On the basis of luminol ECL integrated in FIA, a fiber optic biosensor was developed for the detection of choline [80].

Kusu's group developed an ECL flow-through cell with a carbon fiber electrode employing this cell for the sensitive determination of N-(aminobutyl)-N-ethylisoluminol (ABEI). As a result of this experiment, a straight-line calibration curve for ABEI (r: 0.999) was achieved in the range of 6 fmol to 25 pmol with the detection limit of 6 fmol (S/N = 2). The system was used for the determination of human immunogloblin G (hIgG) with ABEI-labeled anti-hIgG. (Calibration curve of hIgG = 80 pg mL^{-1} to 1.3 ng mL^{-1}, detection limit = 80 pg mL^{-1} (S/N = 2)). Sensitivity and accuracy for sera samples were found to considerably exceed those of the conventional methods, such as single-radial immunodiffusion and npherometric immunoassay. The present system has great potential for immunoassay [81].

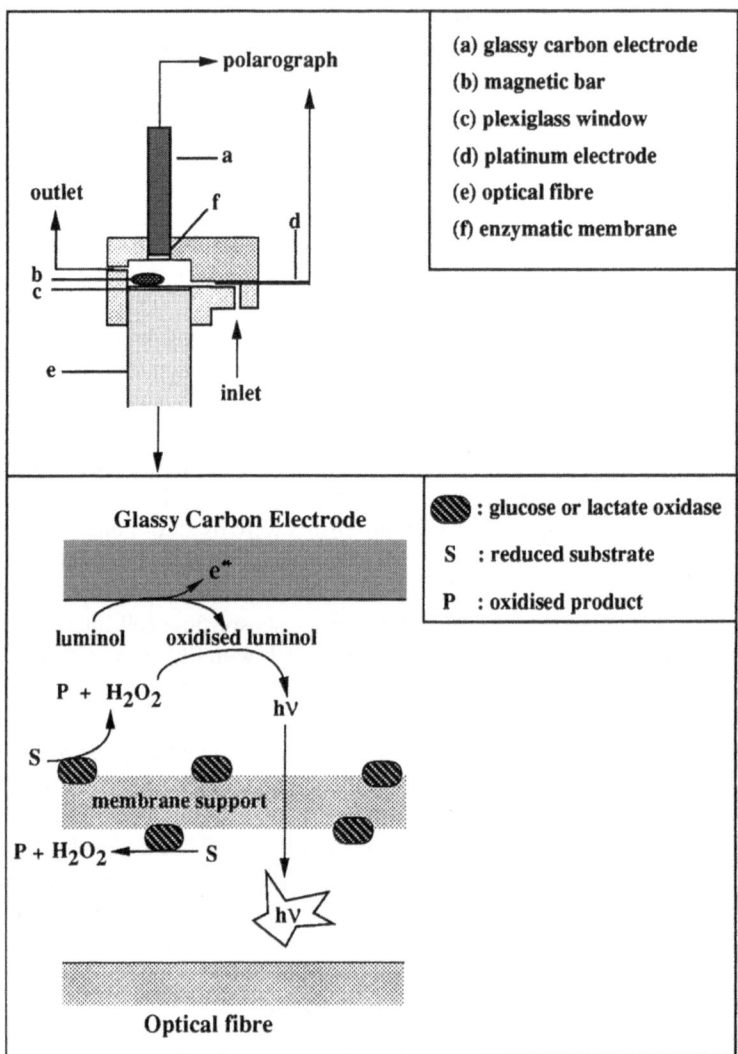

Fig. 5.15 Schematic representation of the flow cell for electrochemiluminescence measurements and detailed representation of the sensor/flow cell interface. Reprinted with permission from Ref. [77]. Copyright 2000 Marcel Dekker, Inc

The use of FI–ECL sensor was also reported for the determination of durabolin in an aqueous system based on CdTe QD films. These QD films, used as a recognizer to determine durabolin, were packed into a homemade cell. The proposed sensor works on the intensive anodic ECL emission which was achieved at a starting potential of +1.3 V (vs. Ag/AgCl) in a carbonate bicarbonate buffer solution with a pH of 9.93 at a CdTe QDs-modified ITO electrode. This approach could easily open new avenues for the applications of QDs in ECL biosensing [82].

H_2O_2 response was explored by another system based on a specially designed facile flow-through cell, employing luminol as the ECL reagent. The developed FIA–ECL system works well with higher sensitivity under optimized experimental conditions, such as the electric parameters, the buffer condition, and the flow rate. Compared with already reported works, the present method offers high sensitivity and lower detection limits. Reactive oxygen species (ROSs) in water vapor is formed during the work of an ultrasonic humidifier with H_2O_2 as index which was monitored by this method. Thus, the proposed system was successfully used to determine the amount of ROSs in some other real samples, such as tap water, drinking water, and river water, was detected with recoveries from 92.0 to 106 % [83].

The flow cell built from solid PTFE, placed in a light-proof steel box was used for the determination of gallic acid. Luminol and KCl solutions were run continuously at 3.5 mL min^{-1} by the peristaltic pump to the mixing valve, which was mixed further with the carrier water stream pumped at 3.5 mL min^{-1} [84]. Sensitive ECL inhibition effects in the presence of adrenaline [85], noradrenaline, and DA were observed for $Ru(bpy)_3^{2+}$/TPA and $Ru(phen)_3^{2+}$/TPA systems. The method was made practical for the determination of commercial pharmaceutical injection samples with satisfied results. Two phenomena, namely potential-dependent ECL and ECL-inhibition, were observed for both $Ru(bpy)_3^{2+}$/TPA and $Ru(phen)_3^{2+}$/TPA systems. It was found that the $Ru(phen)_3^{2+}$/TPA ECL system offers many benefits over $Ru(bpy)_3^{2+}$/TPA system including higher ECL intensity, better reproducibility, and larger inhibition sensitivity to both noradrenaline and DA. Compared with most conventional spectrometry, electrochemical, and HPLC or CE methods in both the sensitivity and LR, the proposed method is very simple, rapid, and sensitive and is possibly applied for trace analyses of biological samples [86].

Catechol, 3,4-dihydroxybenzoic acid, and chlorogenic acid were detected by a FIA-ECL inhibition method of luminol with the detection limit of 1.2×10^{-8}, 2.1×10^{-8}, and 5.2×10^{-9} mol L^{-1}, respectively [87]. In this flow-cell, a sensitive ECL electrode serves as working electrode which is achieved by immobilizing the composite of CdTe QDs (possessed a high quantum yield), CNTs, and chitosan (Chit) on ITO glass. After optimizing parameters such as electrolytic pulse, concentration of TEA, and flow rate, the flow-cell offers the stable ECL. This FIA system detects the sensitive ECL-quenching response of DA within the linear range from 10 pM to 4 nM. The system is practically applied to determine the neurotransmitters in cerebro-spinal fluid (CSF) with DA as the index [88].

The development and application of an electrochemical cell exclusively designed for disposable screen printed carbon electrodes (SPCE) suitable for simultaneous ECL and amperometric detection in sequential injection analysis (SIA) was described by Zen's group. The present flow system is user-friendly, requires less volume, and is easily operatable, since it has facility for PMT via a fiber optic facing the SPCE [89].

Heteropoly acids, the elements comprised in them [90], and perylene (PE) [91] have also been determined by FIA-ECL system. Dong's group explored the effects of heteropoly acids and Triton X-100 on ECL of $Ru(bpy)_3^{2+}$ [90]. Studies on an ECL enzyme immunoassay for TNT (2,4,6-trinitrotoluene) [92] and atrazine [93] were carried out. Antibodies labeled with GO were used in competitive immunoassays where the separation step was performed by concentrating unbound antibodies on the immunosorbent surface. Both of the analytes are detected by H_2O_2 generation as a result of reaction using luminol ECL [92]. An exploded half section through center line of the fountain cell is represented in Fig. 5.16 [93]. Luminol ECL alongwith FIA is used for 2,4-dichlorophenoxyacetic acid (2, 4-D) optical immunosensing employing Glutaraldehyde as a cross-linking agent [94].

In case of Ru(bpy)$_3^{2+}$ ECL integrated with FIA, if analyte and Ru(bpy)$_3^{2+}$ are not immobilized on solid support, they flow continuously and are mixed together at any stage. As analyte concentration in the carrier stream does not remain constant with time, the signal is obtained when the analyte is transported by the carrier into the reaction/detection region. Ru(bpy)$_3^{2+}$ immobilization in poly (p-styrenesulfonate) (PSS)–silica–Triton X-100 composite films [95] and into the Eastman-AQ55D–silica composite thin films [96] was explored alongwith the ECL generated. The modified working electrode was employed for the sensitive and stable ECL detection of oxalate, TPA, NADH, and chlorpromazine (CPZ). Two teflon spacers (thickness: 50 mm) sandwiched between the electrode block and the plexiglas window, create a 24 mL flow cell volume. This volume was determined by evaluating the volume of the cavity between the electrode block and the plexiglas window from the sample inlet to the electrode. Silicone and teflon gaskets were used in order to achieve the following purposes; for example, to protect the teflon spacers, to prevent the rotation of the working electrode, and to avoid leakage of the solution. An Ag/AgCl reference electrode and a counter electrode of Pt wire were connected to the outlet of the flow cell. The flow cell is suggested to have potential applications in ECL and EC detection [95].

Fig. 5.16 Exploded half section through center line of the fountain cell. Adapted from Ref. [93]. Copyright 1997 Elsevier

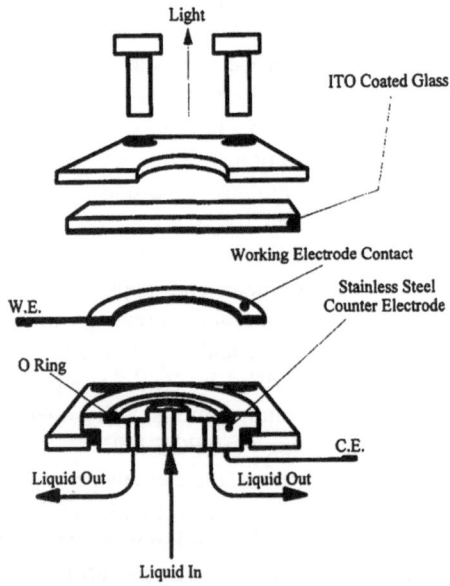

ECL response of allopurinol in aqueous media over a wide pH range (pH 2–13) was studied by Chi et al. [97] They found that allopurinol itself does not have ECL activity but could generate the ECL of $Ru(bpy)_3^{2+}$ in alkaline media giving rise to a sensitive FI–ECL response. Humic acid (HA) was used as modifier to prepare the organic–inorganic hybrid modified glassy carbon electrode based on HA-silica-PVA (poly(vinyl alcohol)) sol–gel composite. Electroactive species of $Ru(bpy)_3^{2+}$ interact with the HA-silica-PVA films to form $Ru(bpy)_3^{2+}$ modified electrodes [98].

Furthermore, trace choline in human urine has been determined by coupling an enzyme reactor with an FI–ECL detector. The enzyme reactor used in this analysis is prepared by covalent immobilization of choline oxidase (ChOx) onto the aminopropyl-controlled pore glass beads (ACPG), which are then carefully packed into a micro column. H_2O_2 produced is catalyzed by the enzyme reactor that is in direct proportion to the concentration of choline. The enzymatically produced H_2O_2 was detected by an ECL detector based on the luminol/H_2O_2 ECL system. Under the optimized condition, LOD of choline as low as 0.05 µM (absolute detection limit was at sub pmol level) was achieved. The method was successfully applied in the determination of choline in the samples of human urine which gave comparable results with those obtained by using the microbore HPLC with an immobilized enzyme reactor-EC detection system [99].

Gallic acid [100] and tetracyclines (TCs) [101] are determined based on an inhibition effect on the $Ru(bpy)_3^{2+}$/TPA, while tannic acid and DA [102] based on its inhibition of the ECL of luminol [103]. A FI-ECL method has been developed for gallic [100] (pH 8.0 phosphate buffer solution) and tannic acid [103]. Quenching effect of the gallic acid on the $Ru(bpy)_3^{2+}$/TPA ECL system was thought as the interaction of electrogenerated $Ru(bpy)_3^{2+*}$ and o-benzoquinone derivative at the electrode surface. This simplistic approach achieves a determination limit of 9.0×10^{-9} mol L^{-1} and a dynamic concentration range of 2×10^{-8} to 2×10^{-5} mol L^{-1} with the relative standard deviation (RSD) of 1.0 % for 1.0×10^{-6} mol L^{-1} gallic acid (n = 11). The method was successfully applicable to the determination of gallic acid in Chinese proprietary medicine— Jianming Yanhou Pian [100]. The method is superior to CL analysis due to its simplistic, rapid, and sensitive approach and no need of catalyst or oxidation agents. The present method has been successfully applied to the determination of tannic acid in real Chinese gall and hop pellet samples [103].

A simple, rapid, and sensitive method for the ECL determination of pyrogallol was found based on potentially dependent inhibition and enhancement effects of pyrogallol on the ECL of luminol in alkaline solutions. The proposed method was successfully applied for the analysis of pyrogallol in a pyrolysis mixture of gallic acid. Pyrogallol on reaction with di-oxygen species generated electrochemically at potentials more positive than 0.50 V versus Ag/AgCl in turn generates superoxide radicals adsorbed on the electrode surface leading to the enhancement of the ECL of luminol [104]. Moreover, a novel flow-through electrolytic cell designed for the determination of captopril [105] and H_2O_2 [106] is reported. Direct CL oxidation of captopril by nascent Mn^{3+}, which was in situ electrogenerated on the near surface of platinum flake electrode by electrochemical oxidizing $MnSO_4$ in sulfuric acid medium is the basis of captopril determination. This method offers a sensitive and rapid approach for the determination of captopril with a detection limit of 8.0×10^{-8} mol L^{-1} and eliminates the drawbacks, which the conventional ECL method has, such as smaller emission area, electrode fouling, etc. [105]. H_2O_2 determination is based on in situ Br_2 production on the near region of a platinum electrode surface by constant current oxidizing KBr in H_2SO_4 medium. In the mean time, a strong CL signal appeared when the basic H_2O_2 solution was injected into the electrolytic cell. On the basis of the above observation, a simple and selective ECL method was introduced for the sensitive determination of H_2O_2 with a detection limit of 3.5×10^{-7} mol L^{-1} [106].

An FI–ECL analysis method in the presence of $Ru(bpy)_3^{2+}$ on ITO glass for the determination of L-lysine, in alkaline Na_2CO_3–$NaHCO_3$ buffer solution, was introduced by Yu's group (Fig. 5.17). A new ECL cell along with this detection method is based on ECL enhancement of $Ru(bpy)_3^{2+}$-L-lysine. This sensor is reusable and shows a significant improvement in sensitivity and selectivity for ECL analysis and has a great potential for analysis of amino acid [107].

It is noteworthy that the enhancing effect of hydrazine on weak ECL signal of the electrooxidation of luminol at a pre-anodized platinum electrode was studied and found stronger than that of hydrazine at a bare platinum electrode [108]. Further studies employ the idea of indirect ion-annihilation ECL for the determination of aromatic tertiary amines. N,N,N',N'-tetramethyl-p-phenylene diamine has been used as model compound. Figure 5.18 represents schematic diagram of the FI manifold and instrumentation used for the electrogeneration of chemiluminescence in this experiment. After optimizing various experimental conditions including applied potentials, electrolyte type and concentration, and flow rate, a detection limit of 4×10^{-7} mol L^{-1} has been obtained with a log-linear range over three decades of concentration [109].

A sensitive and selective FI method for the determination of epinephrine [110], acyclovir (9,2-hydroxyethoxy) methyl guanine [111], allantoin in aqueous alkaline

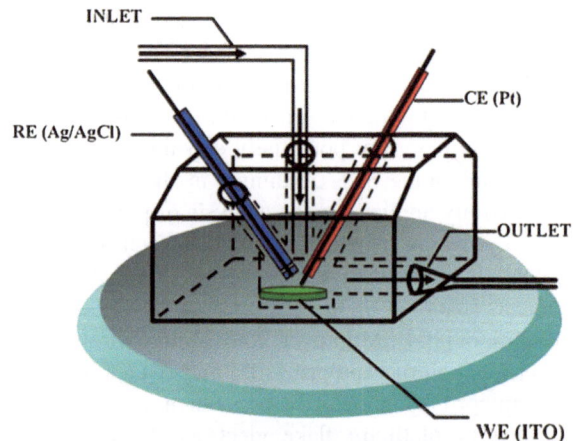

Fig. 5.17 Schematic diagram of the new ECL cell. Reprinted with permission from Ref. [107]. Copyright 2011 The Royal Society of Chemistry

INLET

CE (Pt)

RE (Ag/AgCl)

OUTLET

WE (ITO)

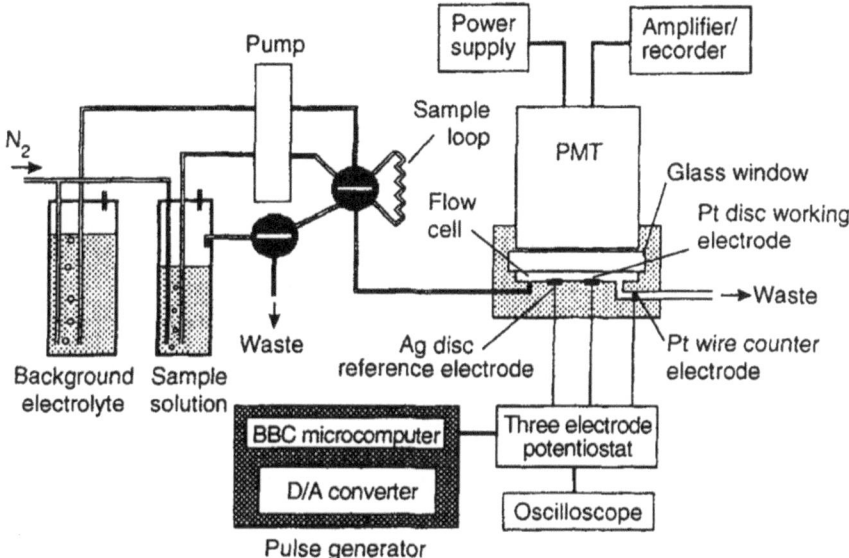

Fig. 5.18 Schematic diagram of the flow injection manifold and instrumentation used for the electrogeneration of chemiluminescence. Reprinted with permission from Ref. [109]. Copyright 1995 The Royal Society of Chemistry

buffer solution (pH 11.0) [112], oxypurinol in alkaline $Ru(bpy)_3^{2+}$ [113], and for the determination of fluoroquinolones based on its sensitizing effect on weak ECL emission of electrochemically oxidized luminol [114] is constructed. ECL batch and continuous-flow measurements of codeine were performed employing a miniaturized and integrated sensor accommodating both transduction electrodes and a photodetector on the same silicon chip (Fig. 5.19). The method was practically used for an analyte exhibiting medical and industrial interests. Codeine with $Ru(bpy)_2(phen)_2$ was found to be more sensitive than other ruthenium complexes, yielding a detection limit equal to 0.1 and 50 mM, in batch and FIA, respectively [115]. The method for the determination of allantoin has advantages over HPLC method in terms of speed and convenience, economics and safe procedure and could be an alternative for places where HPLC equipment is not available [116].

Wang and Yang group in 2003 designed a miniaturized chip-type $Ru(bpy)_3^{2+}$ ECL detection cell for both CE and FIA. The cell was fabricated from two pieces of glass where 0.5-mm-diameter platinum disk was used as working electrode held at +1.15 V (vs. silver wire quasi-reference), the stainless steel guide tubing as counter electrode, and the silver wire as quasi-reference electrode. The device offers advantages of versatility, sensitivity, and accuracy and make it striking for the routine analysis of amine-containing species or oxalate by CE and FI with $Ru(bpy)_3^{2+}$ ECL detection [16].

Fig. 5.19 Scanning
electronic microscope view
of the miniature flow-through
ECL sensor shown before
completion by a plexiglass
top cover plate. Reprinted
with permission from Ref.
[115]. Copyright 1999
Elsevier

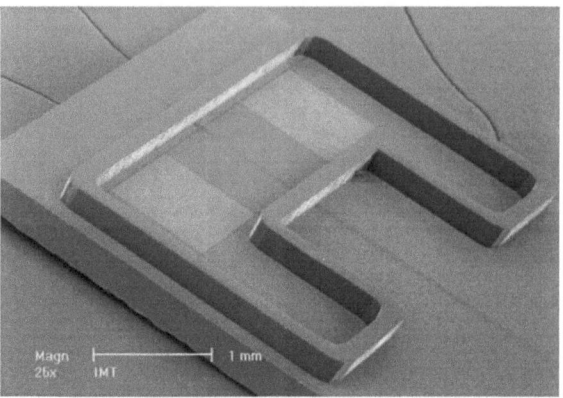

A new cholesterol FIA biosensor based on a luminol/H_2O_2 ECL reaction induced by a glassy carbon electrode polarized at +425 mV versus a Pt pseudo j reference was introduced. The cholesterol biosensor sensing layer used in this method is based on cholesterol oxidase (COD) immobilized on either UltraBind membrane or Immunodyne membrane [117]. Exploration to other literature shows that heroin was determined employing Zeolite Y sieves by immobilizing $Ru(bpy)_3{}^{2+}$ in their supercages. The EC and ECL behaviors of $Ru(bpy)_3{}^{2+}$ immobilized in zeolite Y modified carbon paste electrode was explored. An electrocatalytic response to the oxidation of heroin is displayed on the modified electrode, producing a sensitized ECL signal. The ECL sensor offered good selectivity and long-term stability [118]. The EC behavior of luminol in acidic solution is studied; luminol has the characteristic of formation of polymer on the surface of electrode on electrochemical oxidation which is characterized by CV and the surface enhanced Raman scattering (SERS) spectra. Moreover, enhancement in the ECL of flavin in Na_2CO_3–$NaHCO_3$ buffer solution (pH 9.5) has been found employing this modified electrode. Based on this phenomenon, a very sensitive and selective ECL method for the determination of flavin is presented with a detection limit of 8.3×10^{-8} mol L^{-1} and the linear range response to flavin is 1.0×10^{-7}–1.0×10^{-6} mol L^{-1} [119].

Study of the ECL based on $Ru(bpy)_3{}^{2+}$ revealed that luminescence intensity of monohydric alcohols decreased as alkyl chain length of the molecules increased while increase in the number of hydroxyl groups in a molecule leads to enhancement in luminescence intensity for polyhydric alcohols [120]. Moreover, electrochemical redox potentials, PL, and relative ECL–FIA studies were described for polyamine dendrimers functionalized with electrochemiluminescent polypyridyl Ru(II) complexes, synthesized through the complexation of dendritic polypyridyl ligands to Ru(II) complexes [121]. The adaptability of the newly fabricated thin-layer electrochemical flow cell for amperometric and ECL measurements combined with FI method is demonstrated. This detection is followed by spectrophotometric detection for determination of bromide using the fabricated cell [122].

Ir(III) complexes have attracted considerable interest in ECL detection recently, due to their high emission in various wavelengths. Schematic diagram of a flow cell comprising of three main parts: (a) an electrode body, (b) a PTFE spacer, and (c) a quartz window is shown in Fig. 5.20. Here, ECL from (pq)$_2$Ir(acac) (pq = 2-phenylquinolate, acac = acetylacetonate) is evaluated for the use in FIA. An aqueous solution of the TPA (analyte) and (pq)$_2$Ir(acac) generates ECL by electrochemical initiation after passing through the reaction/observation cell. LOD of 5 nM with a linear range of 3 orders of magnitude in concentration is achieved which is better than that obtained by the conventional Ru(bpy)$_3^{2+}$ system (20 nM). The ECL generation upon various analytes proposes direct applicability of (pq)$_2$Ir(acac) as a post-column detection tool [123].

5.3 ECL Coupled with Solid-Phase Microextraction

Owing to its unique pore network, good strength, and low cost, cement has been used as a new electrode material and solid-phase microextraction material. Cement on mixing with carbon makes a new electrode, cement carbon electrode (CCE) which was employed to reveal the application of cement in SPME by Ru(bpy)$_3^{2+}$ ECL detection of perphenazine (PPZ) up to the detection limit of 3.1×10^{-10} M. Cement-based electrode material may have strong potential to find broad applications in industry related to electrochemistry, such as electrochemical wastewater treatment [124].

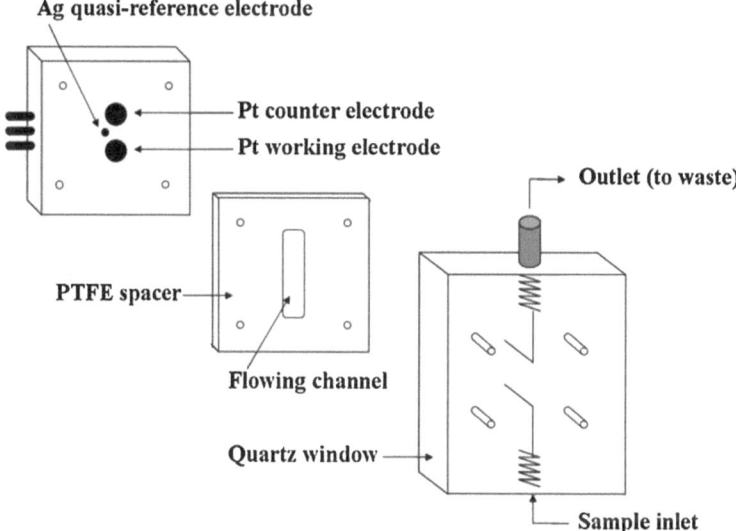

Fig. 5.20 Schematic diagram of a flow cell which comprises of three main parts: (**a**) an electrode body, (**b**) a PTFE spacer, and (**c**) a quartz window. Reprinted with permission from Ref. [123]. Copyright 2011 The Royal Society of Chemistry

L-phenylalanine was determined by a novel FI–ECL method integrated with molecularly imprinted solid-phase extraction (MISPE). In this method, $Ru(bpy)_3^{2+}$ was used as the luminophore and modified indium tin oxide glass as the working electrode. Selective extraction of L-phenylalanine utilized molecularly imprinted polymers, synthesized by self-assembly with functional monomer and crossing linker. The method offered higher sensitivity and repeatability and was applicable in the determination of L-phenylalanine in egg white, chicken, and serum samples. The ECL flow cell employed in this method is equipped with small IR drop, dead volume and flow resistance [125]. Another application of MISPE is quantitative determination of malachite green (MG) residues in fish based on integration with an ECL inhibition method. This strategy is used for the selective extraction and purification of MG and lies on the inhibition effect of MG on ECL on luminol. Under optimized conditions, the quenched ECL intensity was linearly proportional to the logarithm of MG concentration in the range of 20–5,000 ppt with a detection limit of 6 ppt. The method is based on the oxidation reaction with 2,3-dichloro-5,6-dicyano-1,4-benzoquinone (DDQ), which could convert leucomalachite green (LMG) into MG [126].

Determination of melamine in dairy products was performed by developing an ECL enhancement method combined with solid-phase extraction (SPE) in a strong base solution. Scan mode and scan rate of the applied potential, the type of buffer solutions and their pH conditions, were optimized. The presented method has been effectively applied to determine melamine in dairy products including liquid milk, yogurt, and milk powder samples. The detection limit of 0.003 ppb was achieved in the concentration range of 0.01–1.0 ppb [127]. A microscale solid-phase extraction method employing alumina-coated iron oxide nanoparticles ($Fe3O4 @ Al_2O_3$ NPs) as the affinity adsorbent for glyphosate (GLY) and its major metabolite aminomethylphosphonic acid (AMPA) in aqueous solution is described by Whang. $Fe_3O_4 @ Al_2O_3$ NPs are first employed to extract both analytes in 5 mL of aqueous solution. Following extraction/pre-concentration, the analytes were readily desorbed from the NPs by rinsing with $Na_4P_2O_7$ solution and directly analyzed with the derivatization free CE–ECL method. LODs for GLY and AMPA in water were 0.3 and 30 ng mL^{-1}, respectively. The method could be applicable to the analysis GLY in water extract of guava fruit and to the rapid screening of various fruits and crops for residual GLY [128]. Simultaneous determination of three kinds of phenylurea herbicides (PHUs) is performed based on matrix solid-phase dispersion–capillary electrophoresis with electrochemiluminescence detection (MSPD–CE–ECL). Poly-β-cyclodextrin (poly-β-CD) was used as an additive in the running buffer to improve the separation of three analytes (isoproturon, linuron, and diuron). After optimizing the conditions, the three kinds of herbicides were well separated and detected within 8 min with a detection limit of (S/N = 3) of 0.1 µg L^{-1} isoproturon and linuron, and for diuron, a detection limit of 0.2 µg L^{-1} was obtained. The applicability of proposed method was found for the determination of three kinds of herbicides in green vegetable and rice samples with recoveries in the range from 78.1 to 93.8 % [129].

5.4 ECL Coupled with Miscellaneous Techniques

An ECL detection system integrated with an electrically controlled heating cylindrical microelectrode (HME) as the working electrode was developed for the determination of oxalate with a detection limit of 3.0×10^{-4} mol L^{-1} [130]. Steven et al. presented first report on a device that integrates a washing system onto an ELISA plate. The device has an empowered microfluidic handling channel that may be interchanged to allow creation of 4 subsets of the 96-well plates. These four subsets permit the device to measure 4 different target analytes with different secondary antibodies. A hole in the bottom of the well, acting as a "surface tension valve" is used to control flow which keeps the fluid in the wells. To perform ELISA in a point-of-care setting, a microfluidic assay format combined with a CCD sensor is developed. This miniature, simple, and low-cost ELISA-lab-on-a-chip (ELISA–LOC) was designed, fabricated, and tested for the immunological detection of Staphylococcal Enterotoxin B (SEB). The results have demonstrated the feasibility of the proposed protocol in various immunological assays and other complex medical assays without a laboratory [131].

Another group designed a water-soluble functionalized ionic liquid, 1-butyl-3-methylimidazolium dodecyl sulfate $\left([BMIm^+]\left[C_{12}H_{25}SO_4^-\right]\right)$ which was used as supporting electrolyte or as an additive for successful fabrication of a detection method involving MEKC-ECL. 1-butyl-3-methylimidazolium dodecyl sulfate solution used as running buffers greatly enhances ECL intensities and column efficiencies for negative targets as compared to the common supporting electrolytes such as phosphate solution. A little increase in ECL intensity for neutral-charge is also noted, while no change is observed for positive ones due to the electrostatic forces between the large cation $BMIm^+$ and the solutes and the hydrophobic interactions consequencing from the large anion $C_{12}H_{25}SO_4^-$. This novel method will surely exploit novel scopes for MEKC-ECL and functionalized ILs applications in the future [132]. A new approach to collect the ECL signal by coupling the excited state of $Ru(bpy)_3^{2+}$ with the surface plasmons in a thin gold film is proposed (Fig. 5.21). The energy in this approach radiates into the substrate at a defined angle and the surface plasmon-coupled ECL promises to be useful in chemical and biological assays. The exploitation of surface plasmon coupling to collect CL and ECL signals can have potential in a wide range of analytical procedures [133].

It is noteworthy that $Ru(bpy)_3^{2+}$ immobilization on the electrode surface facilitates the enhancement in the ECL intensity [134]. $Ru(bpy)_3^{2+}$ immobilized on an electrode surface comprising sol–gel-derived titania, and Nafion was used for the determination of erythromycin in human urine sample. ECL combined with bipolar electrochemistry has been used for various analytical purposes by Crooks [66, 135–137] and Manz's co-workers [17]. Manz employed ECL with bipolar electrode for the detection of amines and Crooks and co-workers for the detection of $Fe(CN)_6^{3-}$, $Ru(NH_3)_6^{3+}$, benzyl viologen, and various ananlytes. Recently, bipolar electrochemistry has also been coupled to ECL in order to propel light-emitting bipolar electrochemical swimmers Fig. 5.22. In this approach, synergestic action

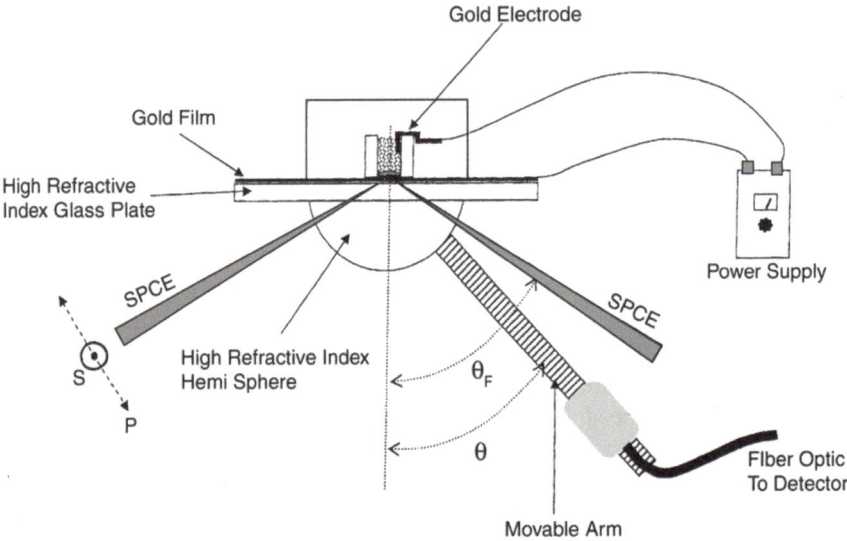

Fig. 5.21 Apparatus used to increase SP–CECL. Reprinted with permission from Ref. [133]. Copyright 2004 Elsevier

Fig. 5.22 Asymmetric light-emitting electrochemical swimmer. Simultaneous reduction of H2O at the cathodic pole (*bottom of the bead*) and oxidation of ECL reagents at the anodic pole (*top of the bead*) induces both motion and light emission from the bead in a glass capillary. P corresponds to a side product of the TPA radicals formed during the ECL process. Reprinted with permission from Ref. [138]. Copyright 2012 Wiley

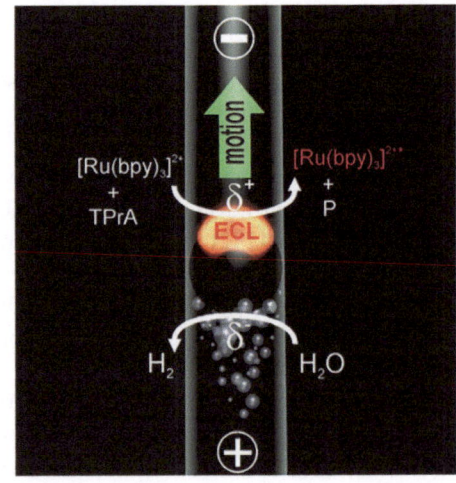

of bipolar electrochemistry with simultaneous bubble production and ECL generation was described. Here, motion on the macroscopic scale occurs concurrently with light emission. Employing this approach ECL offers a method for direct monitoring of the motion of an object, which is extremely valuable in the perspective of autonomous swimmers [138].

References

1. Muzyka EN, Rozhitskii NN (2010) Systems of capillary electrophoresis in electrochemiluminescence analysis. J Anal Chem 65(6):550–564. doi:10.1134/s106193481006002x
2. Harvey N (1929) Luminescence during electrolysis. J Phy Chem 33(10):1456–1459
3. Littig JS, Nieman TA (1992) Quantitation of acridinium esters using electrogenerated chemiluminescence and flow injection. Anal Chem 64(10):1140–1144
4. Zhu L, Li Y, Zhu G (2003) Electrochemiluminescent determination of L-cysteine with a flow-injection analysis system. Anal Sci 19(4):575–578. doi:10.2116/analsci.19.575
5. Chiang MT, Whang CW (2001) Tris(2,2′-bipyridyl)ruthenium(III)-based electrochemiluminescence detector with indium/tin oxide working electrode for capillary electrophoresis. J Chromatogr A 934(1–2):59–66. doi:10.1016/s0021-9673(01)01279-1
6. Dickson JA, Ferris MM, Milofsky RE (1997) Tris (2, 2′-bipyridyl)ruthenium (III) as a chemiluminescent reagent for detection in capillary electrophresis. J High Res Chromatogr 20(12):643–646
7. Wang X, Bobbitt DR (2000) Electrochemically generated $Ru(bpy)_3^{3+}$-based chemiluminescence detection in micellar electrokinetic chromatography. Talanta 53(2):337–345
8. Wang X, Bobbitt DR (1999) In situ cell for electrochemically generated $Ru(bpy)_3^{2+}$-based chemiluminescence detection in capillary electrophoresis. Anal Chim Acta 383(3):213–220
9. Yin XB, Wang E (2005) Capillary electrophoresis coupling with electrochemilurninescence detection: a review. Anal Chim Acta 533(2):113–120. doi:10.1016/j.aca.2004.11.015
10. Du Y, Wang E (2007) Capillary electrophoresis and microchip capillary electrophoresis with electrochemical and electrochemiluminescence detection. J Sep Sci 30(6):875–890. doi:10.1002/jssc.200600472
11. Guo L, Fu F, Chen G (2011) Capillary electrophoresis with electrochemiluminescence detection: fundamental theory, apparatus, and applications. Anal Bioanal Chem 399(10):3323–3343. doi:10.1007/s00216-010-4445-6
12. Lara FJ, Garcia-Campana AM, Ibanez Velasco A (2010) Advances and analytical applications in chemiluminescence coupled to capillary electrophoresis. Electrophoresis 31(12):1998–2027. doi:10.1002/elps.201000031
13. Gilman SD, Silverman CE, Ewing AG (1994) Electrogenerated chemiluminescence detection for capillary electrophoresis. J Microcolumn Sep 6(2):97–106
14. Forbes GA, Nieman TA, Sweedler JV (1997) On-line electrogenerated $Ru(bpy)_3^{3+}$ chemiluminescent detection of Î²-blockers separated with capillary electrophoresis. Anal Chim Acta 347(3):289–293
15. Huang X-J, Wang S-L, Fang Z-L (2002) Combination of flow injection with capillary electrophoresis: 8. Miniaturized capillary electrophoresis system with flow injection sample introduction and electrogenerated chemiluminescence detection. Anal Chim Acta 456(2):167–175
16. Liu JF, Yan JL, Yang XR, Wang EK (2003) Miniaturized tris(2,2′-bipyridyl)ruthenium(II) electrochemiluminescence detection cell for capillary electrophoresis and flow injection analysis. Anal Chem 75(14):3637–3642. doi:10.1021/ac034021y
17. Arora A, Eijkel JCT, Morf WE, Manz A (2001) A wireless electrochemiluminescence detector applied to direct and indirect detection for electrophoresis on a microfabricated glass device. Anal Chem 73(14):3282–3288. doi:10.1021/ac0100300
18. Lim CK (1993) SFY Li, Capillary Electrophoresis—principles, practice and applications, Journal of Chromatography Library, vol 52. Elsevier, Amsterdam, 1992, pp 395, ISBN 044-89433-0. J App Tox 13(4):305–305
19. Chiang MT, Lu MC, Whang CW (2003) A simple and low-cost electrochemiluminescence detector for capillary electrophoresis. Electrophoresis 24(17):3033–3039. doi:10.1002/elps.200305513

20. Chen Y, Lin Z, Sun J, Chen G (2007) A new electrochemiluminescent detection system equipped with an electrically heated carbon paste electrode for CE. Electrophoresis 28(18):3250–3259. doi:10.1002/elps.200700029
21. Cao WD, Yang XR, Wang EK (2004) Determination of reserpine in urine by capillary electrophoresis with electrochemiluminescence detection. Electroanalysis 16(3):169–174. doi:10.1002/elan.200402777
22. Cao WD, Liu JF, Yang XR, Wang E (2002) New technique for capillary electrophoresis directly coupled with end-column electrochemiluminescence detection. Electrophoresis 23(21):3683–3691. doi:10.1002/1522-2683(200211)23
23. Li M, Lee SH (2007) Determination of trimethylamine in fish by capillary electrophoresis with electrogenerated tris(2,2′-bipyridyl)ruthenium(II) chemiluminescence detection. Luminescence 22(6):588–593. doi:10.1002/bio.1006
24. Qiu HB, Yan JL, Sun XH, Liu JF, Cao WD, Yang XR, Wang EK (2003) Microchip capillary electrophoresis with an integrated indium tin oxide electrode-based electrochemiluminescence detector. Anal Chem 75(20):5435–5440. doi:10.1021/ac034500x
25. Chen Y, Lin Z, Chen J, Sun J, Zhang L, Chen G (2007) New capillary electrophoresis-electrochemiluminescence detection system equipped with an electrically heated Ru(bpy)(3)(2+)/multi-wall-carbon-nanotube paste electrode. J Chromatogr A 1172(1): 84–91. doi:10.1016/j.chroma.2007.09.049
26. Hu S, Wang Z-L, Li P-B, Cheng J-K (1997) Amperometric detection in capillary electrophoresis with an etched joint. Anal Chem 69(2):264–267
27. Yin XB, Qiu HB, Sun XH, Yan JL, Liu JF, Wang EK (2004) Capillary electrophoresis coupled with electrochemiluminescence detection using porous etched joint. Anal Chem 76(13):3846–3850. doi:10.1021/ac049743j
28. Yin XB, Kang JZ, Fang LY, Yang XR, Wang EK (2004) Short-capillary electrophoresis with electrochemiluminescence detection using porous etched joint for fast analysis of lidocaine and ofloxacin. J Chromatogr A 1055(1–2):223–228. doi:10.1016/j.chroma.2004.09.001
29. Sreedhar M, Lin YW, Tseng WL, Chang HT (2005) Determination of tertiary amines based on pH junctions and field amplification in capillary electrophoresis with electrochemiluminescence detection. Electrophoresis 26(15):2984–2990. doi:10.1002/elps.200500009
30. Chen GN, Chi YW, Wu XP, Duan JP, Li NB (2003) Chemical oxidation of p-hydroxyphenylpyruvic acid in aqueous solution by capillary electrophoresis with an electrochemiluminescence detection system. Anal Chem 75(23):6602–6607. doi:10.1021/ac034451o
31. Chang PL, Lee KH, Hu CC, Chang HT (2007) CE with sequential light-emitting diodeinduced fluorescence and electro-chemiluminescence detections for the determination of amino acids and alkaloids. Electrophoresis 28(7):1092–1099. doi:10.1002/elps.200600546
32. Zhou M, Ma YJ, Ren XN, Zhou XY, Li L, Chen H (2007) Determination of sinomenine in Sinomenium acutum by capillary electrophoresis with electrochemiluminescence detection. Anal Chim Acta 587(1):104–109. doi:10.1016/j.aca.2007.01.018
33. Sun X, Niu Y, Bi S, Zhang S (2008) Determination of ascorbic acid in individual rat hepatocyte cells based on capillary electrophoresis with electrochemiluminescence detection. Electrophoresis 29(13):2918–2924. doi:10.1002/elps.200700792
34. Sun XH, Liu JF, Cao WD, Yang XR, Wang EK, Fung YS (2002) Capillary electrophoresis with electrochemiluminescence detection of procyclidine in human urine pretreated by ion-exchange cartridge. Anal Chim Acta 470(2):137–145. doi:10.1016/s0003-2670(02)00780-8
35. Sun J, Xu X, Wang C, You T (2008) Analysis of amphetamines in urine with liquid–liquid extraction by capillary electrophoresis with simultaneous electrochemical and electrochemiluminescence detection. Electrophoresis 29(19):3999–4007. doi:10.1002/elps.200700875

36. Huang Y, Pan W, Guo M, Yao S (2007) Capillary electrophoresis with end-column electrochemiluminescence for the analysis of chloroquine phosphate and the study on its interaction with human serum albumin. J Chromatogr A 1154(1–2):373–378. doi:10.1016/j.chroma.2007.02.029

37. Fu Z, Wang L, Li C, Liu Y, Zhou X, Wei W (2009) CE–ECL detection of gatifloxacin in biological fluid after clean-up using SPE. J Sep Sci 32(22):3925–3929. doi:10.1002/jssc.200900508

38. Wang J, Peng Z, Yang J, Wang X, Yang N (2008) Detection of clindamycin by capillary electrophoresis with an end-column electrochemiluminescence of tris(2,2'-bypyridine)ruthenium(II). Talanta 75(3):817–823. doi:10.1016/j.talanta.2007.12.019

39. Deng B, Xie F, Li L, Shi A, Liu Y, Yin H (2010) Determination of galanthamine in Bulbus Lycoridis Radiatae by coupling capillary electrophoresis with end-column electrochemiluminescence detection. J Sep Sci 33(15):2356–2360. doi:10.1002/jssc.201000140

40. Deng B, Ye L, Yin H, Liu Y, Hu S, Li B (2011) Determination of pseudolycorine in the bulb of lycoris radiata by capillary electrophoresis combined with online electrochemiluminescence using ultrasonic-assisted extraction. J Chromatogr B 879(13–14):927–932. doi:10.1016/j.jchromb.2011.03.002

41. Deng B, Kang Y, Li X, Xu Q (2007) Determination of josamycin in rat plasma by capillary electrophoresis coupled with post-column electrochemiluminescence detection. J Chromatogr B 859(1):125–130. doi:10.1016/j.jchromb.2007.09.014

42. Deng B, Shi A, Kang Y, Li L (2011) Determination of metformin hydrochloride using precolumn derivatization with acetaldehyde and capillary electrophoresis coupled with electrochemiluminescence. Lumin: J Biol Chem Lumin 26(6):592–597. doi:10.1002/bio.1276

43. Liu JF, Cao WD, Qiu HB, Sun XH, Yang XR, Wang EK (2002) Determination of sulpiride by capillary electrophoresis with end-column electrogenerated chemiluminescence detection. Cli Chem 48(7):1049–1058

44. Su M, Wei W, Liu S (2011) Analytical applications of the electrochemiluminescence of tris(2, 2'-bipyridyl)ruthenium(II) coupled to capillary/microchip electrophoresis: a review. Anal Chim Acta 704(1–2):16–32. doi:10.1016/j.aca.2011.07.016

45. Johnson TJ, Ross D, Locascio LE (2001) Rapid microfluidic mixing. Anal Chem 74(1):45–51

46. McDonald JC, Duffy DC, Anderson JR, Chiu DT, Wu H, Schueller OJ (2000) Whitesides GM fabrication of microfluidic systems in poly(dimethylsiloxane). Electrophoresis 21(1):27–40

47. McDonald JC, Whitesides GM (2002) Poly(dimethylsiloxane) as a material for fabricating microfluidic devices. Acc Chem Res 35(7):491–499

48. Effenhauser CS, Bruin GJM, Paulus A, Ehrat M (1997) Integrated capillary electrophoresis on flexible silicone microdevices: analysis of DNA restriction fragments and detection of single DNA molecules on microchips. Anal Chem 69(17):3451–3457

49. Duffy DC, McDonald JC, Schueller OJA, Whitesides GM (1998) Rapid prototyping of microfluidic systems in poly(dimethylsiloxane). Anal Chem 70(23):4974–4984

50. Zhan W, Alvarez J, Crooks RM (2002) A two-channel microfluidic sensor that uses anodic electrogenerated chemiluminescence as a photonic reporter of cathodic redox reactions. Anal Chem 75(2):313–318

51. Zhan W, Alvarez J, Sun L, Crooks RM (2003) A multichannel microfluidic sensor that detects anodic redox reactions indirectly using anodic electrogenerated chemiluminescence. Anal Chem 75(6):1233–1238

52. Reyes DR, Iossifidis D, Auroux P-A, Manz A (2002) Micro total analysis systems. 1. Introduction, theory, and technology. Anal Chem 74(12):2623–2636

53. Auroux P-A, Iossifidis D, Reyes DR, Manz A (2002) Micro total analysis systems. 2. Analytical standard operations and applications. Anal Chem 74(12):2637–2652

54. Zhan W, Alvarez J, Crooks RM (2002) Electrochemical sensing in microfluidic systems using electrogenerated chemiluminescence as a photonic reporter of redox reactions. J Am Chem Soc 124(44):13265–13270
55. Cao W, Jia J, Yang X, Dong S, Wang E (2002) Capillary electrophoresis with solid-state electrochemiluminescence detector. Electrophoresis 23(21):3692–3698
56. Du Y, Wang E (2008) Separation and detection of narcotic drugs on a microchip using micellar electrokinetic chromatography and electrochemiluminescence. Electroanalysis 20(6):643–647. doi:10.1002/elan.200704117
57. Zhao XC, You TY, Qiu HB, Yan JL, Yang XR, Wang EK (2004) Electrochemiluminescence detection with integrated indium tin oxide electrode on electrophoretic microchip for direct bioanalysis of lincomycin in the urine. J Chromatogr B 810(1):137–142. doi:10.1016/j.jchromb.2004.07.018
58. Yin XB, Du Y, Yang XR, Wang EK (2005) Microfluidic chip with electrochemiluminescence detection using 2-(2-aminoethyl)-1-methylpyrrolidine labeling. J Chromatogr A 1091(1–2):158–162. doi:10.1016/j.chroma.2005.07.046
59. Qiu HB, Yin XB, Yan JL, Zhao XC, Yang XR, Wang EK (2005) Simultaneous electrochemical and electrochemiluminescence detection for microchip and conventional capillary electrophoresis. Electrophoresis 26(3):687–693. doi:10.1002/elps.200410015
60. Du Y, Wei H, Kang JZ, Yan JL, Yin XB, Yang XR, Wang EK (2005) Microchip capillary electrophoresis with solid-state electrochemiluminescence detector. Anal Chem 77(24):7993–7997. doi:10.1021/ac051369f
61. Ding S-N, Xu J-J, Zhang W-J, Chen H-Y (2006) Tris(2, 2'-bipyridyl)ruthenium(II)-zirconia-nafion composite modified electrode applied as solid-state electrochemiluminescence detector on electrophoretic microchip for detection of pharmaceuticals of tramadol, lidocaine and ofloxacin. Talanta 70(3):572–577. doi: 10.1016/j.talanta.2006.01.017
62. Ding SN, Xu JJ, Chen HY (2006) Microchip capillary electrophoresis coupled with an end-column electrochemiluminescence detection. Talanta 70(2):403–407. doi:101016/jtalanta200602063
63. Yang H, X-c Li, Yang F, Feng J, M-y Lin, Z-g Chen (2011) Electrochemiluminescence detection system for microchip capillary electrophoresis and its application to pharmaceutical analysis. Microchim Acta 175(1–2):193–199. doi:10.1007/s00604-011-0670-8
64. Niskanen AJ, Ylinen-Hinkka T, Kulmala S, Franssila S (2011) Integrated microelectrode hot electron electrochemiluminescent sensor for microfluidic applications. Sens and Actuators B-Chem 152(1):56–62. doi:10.1016/j.snb.2010.09.049
65. Yuan B, Huang J, Sun J, You T (2009) A novel technique for NACE coupled with simultaneous electrochemiluminescence and electrochemical detection for fast analysis of tertiary amines. Electrophoresis 30(3):479–486. doi:10.1002/elps.200800253
66. Zhan W, Alvarez J, Crooks RM (2003) A two-channel microfluidic sensor that uses anodic electrogenerated chemiluminescence as a photonic reporter of cathodic redox reactions. Anal Chem 75(2):313–318. doi:10.1021/ac020488h
67. Yan JL, Liu JF, Cao WD, Sun XH, Yang XR, Wang EK (2004) Determination of benzhexol hydrochloride by capillary zone electrophoresis with an end-column electrochemiluminescence detection. Microchem J 76(1–2):11–16. doi:10.1016/j.microc.2003.10.003
68. Wang J, Zhang X, Pi F, Wang X, Yang N (2009) Tris(2,2'-bipyridyl) ruthenium(II)-bisoprolol-based electrochemiluminescence coupled with capillary zone electrophoresis. Electrochim Acta 54(8):2379–2384. doi:10.1016/j.electacta.2008.10.066
69. Li H, Liu X, Niu W, Zhu S, Fan L, Shi L, Xu G (2008) CEC with tris(2,2'-bipyridyl) ruthenium(II) electrochemiluminescent detection. Electrophoresis 29(22):4475–4481. doi:10.1002/elps.200800088

70. Yin X-B, Guo J-M, Wei W (2010) Dual-cloud point extraction and tertiary amine labeling for selective and sensitive capillary electrophoresis-electrochemiluminescent detection of auxins. J Chromatogr A 1217(8):1399–1406. doi:10.1016/j.chroma.2009.12.029

71. Liu YJ, Pan W, Liu Q, Yao SZ (2005) Study on the enhancement of Ru(bpy)$_3^{2+}$ electrochemiluminescence by nanogold and its application for pentoxyverine detection. Electrophoresis 26(23):4468–4477. doi:10.1002/elps.200500391

72. Li JG, Yan QY, Gao YL, Ju HX (2006) Electrogenerated chemiluminescence detection of amino acids based on precolumn derivatization coupled with capillary electrophoresis separation. Anal Chem 78(8):2694–2699. doi:10.1021/ac052092m

73. Li X, Zhu D, You T (2011) Simultaneous analysis of six cardiovascular drugs by capillary electrophoresis coupled with electrochemical and electrochemiluminescence detection, using a chemometrical optimization approach. Electrophoresis 32(16):2139–2147. doi:10.1002/elps.201100074

74. Zhu L, Li YX, Zhu GY (2002) A novel flow through optical fiber biosensor for glucose based on luminol electrochemiluminescence. Sens Actuators, B 86(2–3):209–214. doi:10.1016/s0925-4005(02)00173-9

75. Lei R, Wang X, Zhu S, Li N (2011) A novel electrochemiluminescence glucose biosensor based on alcohol-free mesoporous molecular sieve silica modified electrode. Sens Actuators, B 158(1):124–129. doi:10.1016/j.snb.2011.05.054

76. Kremeskotter J, Wilson R, Schiffrin DJ, Luff BJ, Wilkinson JS (1995) Detection of glucose via electrochemiluminescence in a thin-layer cell with a planar optical wave-guide. Meas Sci Tech 6(9):1325–1328. doi:10.1088/0957-0233/6/9/012

77. Marquette CA, Blum LJ (1999) Luminol electrochemiluminescence-based fibre optic biosensors for flow injection analysis of glucose and lactate in natural samples. Anal Chim Acta 381(1):1–10. doi:10.1016/s0003-2670(98)00703-x

78. Marquette CA, Degiuli A, Blum LJ (2000) Fiberoptic biosensors based on chemiluminescent reactions. App Biochem and Biotech 89(2–3):107–115. doi:10.1385/abab:89:2-3:107

79. Wang CY, Huang HJ (2003) Flow injection analysis of glucose based on its inhibition of electrochemiluminescence in a Ru(bpy)$_3^{2+}$-tripropylamine system. Anal Chim Acta 498(1–2):61–68. doi:10.1016/j.aca.2003.08.064

80. Tsafack VC, Marquette CA, Leca B, Blum LJ (1999) An electrochemiluminescence-based fibre optic biosensor for choline flow injection analysis. Analyst 125(1):151–155. doi:10.1039/a907709j

81. Arai K, Takahashi K, Kusu F (1999) An electrochemiluminescence flow through-cell and its applications to sensitive immunoassay using N-(aminobutyl)-N-ethylisoluminol. Anal Chem 71(11):2237–2240. doi:10.1021/ac9810361

82. Wan F, Yu J, Yang P, Ge S, Yan M (2011) An electrochemiluminescence sensor for determination of durabolin based on CdTe QD films by layer-by-layer self-assembly. Anal Bioanal Chem 400(3):807–814. doi:10.1007/s00216-011-4808-7

83. Chen M, Wei X, Tu Y (2011) A luminol-based micro-flow-injection electrochemiluminescent system to determine reactive oxygen species. Talanta 85(3):1304–1309. doi:10.1016/j.talanta.2011.06.002

84. Sun YG, Cui H, Li YH, Li SF, Lin XQ (2000) Determination of gallic acid by flow injection with electrochemiluminescent detection. Anal Lett 33(15):3239–3252

85. Li F, Cui H, Lin XQ (2002) Determination of adrenaline by using inhibited Ru(bpy)$_3^{2+}$ electrochemiluminescence. Anal Chim Acta 471(2):187–194. doi:10.1016/s0003-2670(02)00930-3

86. Li F, Pang YQ, Lin XQ, Cui H (2003) Determination of noradrenaline and dopamine in pharmaceutical injection samples by inhibition flow injection electrochemiluminescence of ruthenium complexes. Talanta 59(3):627–636. doi:10.1016/s0039-9140(02)00576-3

87. Sun YG, Cui H, Li YH, Lin XQ (2000) Determination of some catechol derivatives by a flow injection electrochemiluminescent inhibition method. Talanta 53(3):661–666. doi:10.1016/s0039-9140(00)00550-6

88. Zhao J, Chen M, Yu C, Tu Y (2011) Development and application of an electrochemiluminescent flow-injection cell based on CdTe quantum dots modified electrode for high sensitive determination of dopamine. Analyst 136(19):4070–4074. doi:10.1039/c1an15458c

89. Chin M-H, Wu H, Chen J-C, Muthuraman G, Zen J-M (2007) Disposable screen-printed carbon electrodes for dual electrochemiluminescence/amperometric detection: Sequential injection analysis of oxalate. Electroanalysis 19(22):2301–2306. doi:10.1002/elan.200703984

90. Xu GB, Cheng L, Dong SJ (1999) Effects of heteropoly acids and surfactant on electrochemiluminescence of tris(2,2′-bipyridine) ruthenium(II). Anal Lett 32(11):2311–2326. doi:10.1080/00032719908542972

91. Itagaki M, Kobari N, Watanabe K (2004) Electrochemiluminescence impedance of perylene in acetonitrile. J Electroanal Chem 572(2):329–333. doi:10.1016/j.jelechem.2003.12.038

92. Wilson R, Clavering C, Hutchinson A (2003) Electrochemiluminescence enzyme immunoassay for TNT. Analyst 128(5):480–485. doi:10.1039/b301942j

93. Wilson R, Barker MH, Schiffrin DJ, Abuknesha R (1997) Electrochemiluminescence flow injection immunoassay for atrazine. Biosens Bioelectron 12(4):277–286. doi:10.1016/s0956-5663(96)00067-x

94. Marquette CA, Blum LJ (1998) Electrochemiluminescence of luminol for 2,4-D optical immunosensing in a flow injection analysis system. Sens Actuators, B 51(1–3):100–106. doi:10.1016/s0925-4005(98)00175-0

95. Wang HY, Xu GB, Dong SJ (2001) Electrochemiluminescence of tris(2,2′-bipyridine)ruthenium(II) immobilized in poly(p-styrenesulfonate)-silica-Triton X-100 composite thin-films. Analyst 126(7):1095–1099. doi:10.1039/b100376n

96. Wang HY, Xu GB, Dong SJ (2003) Electrochemiluminescence sensor using tris(2,2′-bipyridyl)ruthenium(II) immobilized in Eastman-AQ55D-silica composite thin-films. Anal Chim Acta 480(2):285–290. doi:10.1016/s0003-2670(03)00049-7

97. Chi YW, Xie JL, Chen GN (2006) Electrochemiluminescent behavior of allopurinol in the presence of Ru(bpy)$_3{}^{2+}$. Talanta 68(5):1544–1549. doi:10.1016/j.talanta.2005.08.013

98. Ding SN, Xu JJ, Chen HY (2005) Electrogenerated chemiluminescence of tris(2,2′-bipyridyl) ruthenium(II) immobilized in humic acid-silica-poly(vinyl alcohol) composite films. Electroanal 17(17):1517–1522. doi:10.1002/elan.200403249

99. Jin J, Muroga M, Takahashi F, Nakamura T (2010) Enzymatic flow injection method for rapid determination of choline in urine with electrochemiluminescence detection. Bioelectrochem 79(1):147–151. doi:10.1016/j.bioelechem.2009.12.005

100. Lin XQ, Li F, Pang YQ, Cui H (2004) Flow injection analysis of gallic acid with inhibited electrochemiluminescence detection. Anal Bioanal Chem 378(8):2028–2033. doi:10.1007/s00216-004-2519-z

101. Pang YQ, Cui H, Zheng HS, Wan GH, Liu LJ, Yu XF (2005) Flow injection analysis of tetracyclines using inhibited Ru(bPY)$_3{}^{2+}$/tripropylamine electrochemiluminescence system. Luminescence 20(1):8–15. doi:10.1002/bio.793

102. Zhu LD, Li YX, Zhu GY (2002) Flow injection determination of dopamine based on inhibited electrochemiluminescence of luminol. Anal Lett 35(15):2527–2537. doi:10.1081/al-120016542

103. Sun YG, Cui H, Li YH, Zhao HZ, Lin XQ (2000) Flow injection analysis of tannic acid with inhibited electrochemiluminescent detection. Anal Lett 33(11):2281–2291. doi:10.1080/00032710008543189

104. Sun YG, Cui H, Lin XQ, Li YH, Zhao HZ (2000) Flow injection analysis of pyrogallol with enhanced electrochemiluminescent detection. Anal Chim Acta 423(2):247–253. doi:10.1016/s0003-2670(00)01121-1

105. Zheng XW, Zhang ZJ, Li BX (2001) Flow injection chemiluminescence determination of captopril with in situ electrogenerated Mn^{3+} as the oxidant. Electroanalysis 13(12):1046–1050. doi:10.1002/1521-4109(200108)13

106. Zheng XW, Yang M, Zhang ZJ (1999) Flow injection chemiluminescence determination of hydrogen peroxide with in situ electrogenerated Br^{-2} as the oxidant. Anal Lett 32(15):3013–3028

107. Wang S, Yu J, Wan F, Ge S, Yan M, Zhang M (2011) Flow injection electrochemiluminescence determination of L-lysine using tris (2,2′-bipyridyl) ruthenium(II) $(Ru(bpy)_3^{2+}$ on indium tin oxide (ITO) glass. Anal Meth 3(5):1163–1167. doi:10.1039/c0ay00632g

108. Zheng XW, Zhang ZJ, Guo ZH, Wang Q (2002) Flow-injection electrogenerated chemiluminescence detection of hydrazine based on its in situ electrochemical modification at a pre-anodized platinum electrode. Analyst 127(10):1375–1379. doi:10.1039/b203172h

109. Knight AW, Greenway GM (1995) Indirect, ion-annihilation electrogenerated chemiluminescence and its application to the determination of aromatic tertiary-amines. Analyst 120(4):1077–1082. doi:10.1039/an9952001077

110. Zheng XW, Guo ZH, Zhang ZJ (2001) Flow-injection electrogenerated chemiluminescence determination of epinephrine using luminol. Anal Chim Acta 441(1):81–86. doi:10.1016/s0003-2670(01)01090-x

111. Lv J, Luo LR, Zhang ZJ (2004) On-line galvanic cell generated electrochemiluminescence determination of acyclovir based on the flow injection sampling. Anal Chim Acta 510(1):35–39. doi:10.1016/j.aca.2003.12.056

112. Chen X, Tao Y, Zhao L, Xie ZH, Chen GN (2005) Preliminary electrochemiluminescence study of allantoin in the presence of tris(2, 2′-bipyridine) ruthenium(II). Luminescence 20(3):109–116. doi:10.1002/bio.828

113. Chi Y, Dong Y, Chen G (2007) Investigation on the electrochemiluminescent behaviors of oxypurinol in alkaline Ru(bpy) $_3^{2+}$ solution using a flow injection analytical system. Electrochem Comm 9(4):577–583. doi:10.1016/j.elecom.2006.09.030

114. Ma HY, Zheng XW, Zhang ZJ (2005) Flow-injection electrogenerated chemiluminescence determination of fluoroquinolones based on its sensitizing effect. Luminescence 20(4–5):303–306. doi:10.1002/bio.838

115. Michel PE, Fiaccabrino GC, de Rooij NF, Koudelka-Hep M (1999) Integrated sensor for continuous flow electrochemiluminescent measurements of codeine with different ruthenium complexes. Anal Chim Acta 392(2–3):95–103

116. Zhao C, Chai X, Tao S, Li M, Jiao K (2008) Selective determination of diphenhydramine in compound pharmaceutical containing ephedrine by flow-injection electrochemiluminescence. Anal Sci 24(4):535–538. doi:10.2116/analsci.24.535

117. Marquette CA, Ravaud S, Blum LJ (2000) Luminol electrochemiluminescence-based biosensor for total cholesterol determination in natural samples. Anal Lett 33(9):1779–1796. doi:10.1080/00032710008543158

118. Zhuang YF, Zhang DM, Ju HX (2005) Sensitive determination of heroin based on electrogenerated chemiluminescence of tris(2, 2′-bipyridyl)ruthenium(II) immobilized in zeolite Y modified carbon paste electrode. Analyst 130(4):534–540. doi:10.1039/b415430d

119. Zhang GF, Chen HY (2000) Studies of polyluminol modified electrode and its application in electrochemiluminescence analysis with flow system. Anal Chim Acta 419(1):25–31. doi:10.1016/s0003-2670(00)00983-1

120. Chen X, Sato M, Lin YJ (1998) Study of the electrochemiluminescence based on tris(2,2′-bipyridine) ruthenium(II) and alcohols in a flow injection system. Microchem J 58(1):13–20. doi:10.1006/mchj.1997.1503

121. Lee DN, Park HS, Kim EH, Jun YM, Lee JY, Lee WY, Kim BH (2006) Synthesis of novel electrochemfluminescent polyamine dendrimers functionalized with polypyridyl Ru(II) complexes and their electrochemical properties. Bull Kor Chem Soc 27(1):99–105

122. Haghighi B, Aghajari H, Bozorgzadeh S, Gorton L (2011) Fabrication and characterization of a thin-layer electrochemical flow cell and its application for flow analysis. Anal Lett 44(1–3):258–270. doi:10.1080/00032719.2010.500763

123. Shin I-S, Kang Y-T, Lee J-K, Kim H, Kim TH, Kim JS (2011) Evaluation of electrogenerated chemiluminescence from a neutral Ir(III) complex for quantitative analysis in flowing streams. Analyst 136(10):2151–2155. doi:10.1039/c1an15045f

124. Yuan Y, Li H, Han S, Hu L, Xu G (2010) Application of cement as new electrode material and solid-phase microextraction material demonstrated by electrochemiluminescent detection of perphenazine. Talanta 84(1):49–52

125. Lu J, Ge S, Wan F, Yu J (2012) Detection of L-phenylalanine using molecularly imprinted solid-phase extraction and flow injection electrochemiluminescence. J Sep Sci 35(2):320–326. doi:10.1002/jssc.201100787

126. Guo Z, Gai P, Hao T, Duan J, Wang S (2011) Determination of malachite green residues in fish using a highly sensitive electrochemiluminescence method combined with molecularly imprinted solid phase extraction. J Agr Food Chem 59(10):5257–5262. doi:10.1021/jf2008502

127. Guo Z, Gai P, Hao T, Wang S, Wei D, Gan N (2011) Determination of melamine in dairy products by an electrochemiluminescent method combined with solid-phase extraction. Talanta 83(5):1736–1741. doi:10.1016/j.talanta.2010.12.013

128. Hsu C–C, Whang C-W (2009) Microscale solid phase extraction of glyphosate and aminomethylphosphonic acid in water and guava fruit extract using alumina-coated iron oxide nanoparticles followed by capillary electrophoresis and electrochemiluminescence detection. J Chromatogr A 1216(49):8575–8580. doi:10.1016/j.chroma.2009.10.023

129. Wang Y, Xiao L, Cheng M (2011) Determination of phenylureas herbicides in food stuffs based on matrix solid-phase dispersion extraction and capillary electrophoresis with electrochemiluminescence detection. J Chromatogr A 1218(50):9115–9119. doi:10.1016/j.chroma.2011.10.048

130. Lin ZY, Sun JJ, Chen JH, Guo L, Chen GN (2006) A new electrochemiluminescent detection system equipped with an electrically controlled heating cylindrical microelectrode. Anal Chim Acta 564(2):226–230. doi:10.1016/j.aca.2006.01.095

131. Sun S, Yang M, Kostov Y, Rasooly A (2010) ELISA-LOC: lab-on-a-chip for enzyme-linked immunodetection. Lab Chip 10(16):2093–2100. doi:10.1039/c003994b

132. Xu Y, Fang L, Wang E (2009) Successful establishment of MEKC with electrochemiluminescence detection based on one functionalized ionic liquid. Electrophoresis 30(2):365–371. doi:101002/elps200800187

133. Zhang J, Gryczynski Z, Lakowicz JR (2004) First observation of surface plasmon-coupled electrochemiluminescence. Chem Phy Lett 393(4–6):483–487. doi:10.1016/j.cplett.2004.06.050

134. Choi HN, Cho SH, Lee WY (2003) Electrogenerated chemiluminescence from tris(2,2′-bipyridyl)ruthenium(II) immobilized in titania-perfluorosulfonated ionomer composite films. Anal Chem 75(16):4250–4256

135. Mavreĺ F, Anand RK, Laws DR, Chow KF, Chang BY, Crooks JA, Crooks RM (2010) Bipolar Electrodes: a useful tool for concentration, separation, and detection of analytes in microelectrochemical systems. Anal Chem 82(21):8766–8774

136. Chow K-F, Mavreĺ F, Crooks RM (2008) Wireless electrochemical DNA microarray sensor. J Am Chem Soc 130(24):7544–7545

137. Chow K-F, Mavreĺ F, Crooks JA, Chang BY, Crooks RM (2009) A large-scale, wireless electrochemical bipolar electrode microarray. J Am Chem Soc 131(24):8364–8365

138. Sentic M, Loget G, Manojlovic D, Kuhn A, Sojic N (2012) Light-emitting electrochemical "Swimmers". Angew Chem Int Ed 51(45):11284–11288

Chapter 6
Quenching of ECL

Abstract ECL quenching may play an important role in designing new methodologies for sensitive detection of analytes. Quenching proposes prospective advantages in the framework of ECL and has acquired considerable attention and is inextricably associated with the selectivity of luminophore and co-reactant. It can be used in diverse fields in the detection of many analytes, DNA detection, and hybridization, etc. Processes, reactions, and equations involving quenching are discussed as well.

Keywords Quenching · Förster transfer · Stern–Volmer equation · Electron-transfer-quenching path · Cathodic ECL · Energy scavenging process

Quenching can occur mainly by two ways, namely energy transfer or electron transfer. The energy transfer, sometimes called Förster transfer, occurs at short distances between the reactants, by a direct electrodynamic interaction between D* and Q (Eq. 6.1). It is favored by the electronic energy of D* being greater than that of Q* and a large overlap of the emission band of D* with the absorption band of Q. An excited state quenched by another molecule (a quencher, Q) to produce the ground state can be shown as

$$D^* + Q \rightarrow D + Q^* \tag{6.1}$$

where Q* can often decay to the ground state without emission.

Electron-transfer quenching is related to the excitation energy of the molecule, as excited state is easier to oxidize and easier to reduce than the corresponding ground state of the same molecule. If this amount of energy is essentially equal to the amount of excitation energy of the molecule, electron-transfer quenching occurs. Thus, if the reduction in A, (Eq. 6.2), occurs with E° $(A, A^{-\bullet})$, the reduction in A* occurs at E° $(A, A^{-\bullet})$ $-E^*$, where E* is the excitation energy of A and E° $(A, A^{-\bullet})$ is the standard electrode potential [1].

$$A + e \leftrightarrows A^{-\bullet} \tag{6.2}$$

S. Parveen et al., *Electrogenerated Chemiluminescence*,
SpringerBriefs in Molecular Science, DOI: 10.1007/978-3-642-39555-0_6,
© The Author(s) 2013

For example, if the ground state is reduced at -1.0 V and the excitation energy is 2.5 eV, the excited state would be reduced at $+1.5$ V. Similarly, oxidation of the excited state occurs at $E°$ $(D^{+\bullet}, D)$ $-E^*$. Thus, excited states can be quenched rather easily by an electron-transfer process, as shown in Fig. 6.1, so that the radical ions are effective quenchers.

The kinetics of a quenching reaction, (Eq. 6.1), are governed by the Stern–Volmer equation (Eq. 6.3)

$$\phi^0/\phi - 1 = R^0/R - 1 = \kappa_Q \tau_0 [Q] \qquad (6.3)$$

where ϕ^0 and ϕ are the fluorescence efficiencies and R^0 and R are the fluorescence responses in the absence and presence of a quencher at concentration $[Q]$, respectively, k_Q is the rate constant for quenching, and τ_0 is the lifetime of the excited state in the absence of a quencher [1]. Metal electrodes can similarly act as quenchers by either energy-transfer process or electron-transfer process. The energy-transfer mode is analogous to Förster transfer [2, 3], while electron-transfer process follows the same opinion as those given above for redox quenching [4]. The effective quenching by metals and semiconductor electrodes has also been reported.

Quenching offers potential advantages in the context of ECL, has garnered considerable attention, and is inextricably linked to the selectivity of luminophore and co-reactant. It can be used in a variety of fields in the detection of many analytes, DNA detection, and hybridization, etc. Phenolic quenching of the $Ru(bpy)_3^{2+}$ ECL reaction was first reported in a paper describing the determination of codeine and structurally related opiate alkaloids of pharmaceutical importance [5]. More recently, quenching studies [6, 7] have offered complementary reaction schemes invoking phenols as scavengers of radical intermediates operating within the

Fig. 6.1 Molecular orbital representation of electron-transfer quenching. Reprinted with permission from Ref. [1]. Copyright 2004 Taylor & Francis

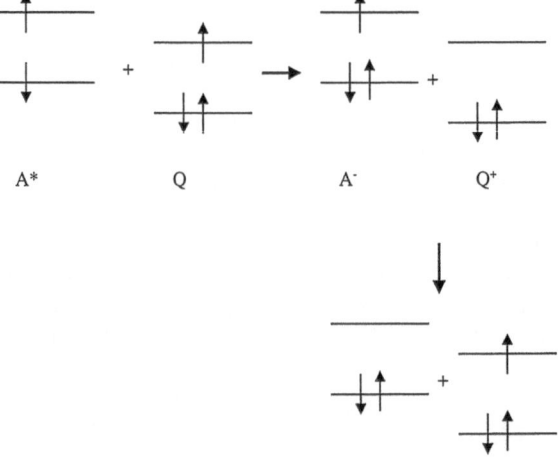

excitation pathways. In 1992, Hoffman and co-workers [8] described the quenching of excited $Cr(bpy)_3^{3+}$ by phenol. Aromatic compounds (such as benzene) have also been found to quench chemiluminescence when attached directly to certain groups [9]. Later, McCall et al. [10] and Richter group [11] observed the inhibition of $Ru(bpy)_3^{2+}$ ECL by phenols, hydroquinone, catechol, and benzoquinones.

$Ru(bpy)_3^{2+}$/TEA ECL has the ability to be inhibited strongly by many compounds with electrochemical oxidation activities, including inorganic and organic compounds. Inorganic compounds include NO^{2-}, NO, $[Fe(CN)_6]^{4-}$, I^-, and H_2O_2, and organic compounds may contain phenol-containing compounds (phenol, tyrosine, catechol, hydroquinone, homogentisic acid), indoles (indole, indole-3-acetic acid), aniline, imidazoles (imidazole, methimazole), purines (xanthine, hypoxanthine, adenine, guanine, allopurinol, oxypurinol), and pyrimidines (thymine, cytosine, uracil, thiouracil). On the basis of these electroactive inhibitors, new electrochemical oxidation-inhibiting mechanism has been proposed explaining the inhibition of $Ru(bpy)_3^{2+}$/TEA ECL [12]. Inorganic oxidants also create inhibited $Ru(bpy)_3^{2+}$ ECL, investigated by Qiu. It was found that a number of inorganic oxidants can quench the ECL of $Ru(bpy)_3^{2+}$/TPA system, and the logarithm of the decrease in ECL intensity was proportional to the logarithm of analyte concentrations. Based on which, a sensitive approach for detection of these inorganic oxidants was established [13].

Two kinds of reactions happened in a co-reactant-based ECL process. One was related to ECL emission, while the other to ECL inhibition. When the reactions associated with ECL inhibition were dominating, the total ECL emission would be decreased by increasing the co-reactant concentration. Guo systematically investigated the ECL inhibition behavior between co-reactants for the first time. The results demonstrated that ECL inhibition did not only happen within the same type of co-reactants, but also between two different types of co-reactants. Mostly, it is envisioned to develop sensitive inhibited ECL detection methods for weak co-reactants instead of ECL enhancement. The proposed ECL inhibition mechanism shows the consumption of co-reactant intermediates without light emission [14]. Construction of DNA molecular logic gates related to ECL generation is introduced by Li based on the T-rich or C-rich oligonucleotides for the selective analysis of Hg^{2+} and Ag^+ ions using energy- or electron-transfer-quenching path [15].

Recently, the ECL of $Ru(bpy)_3^{2+}$ and the quenching of ECL by naphthol using RTIL, 1-butyl-3-methylimidazolium tetrafluoroborate [Bmim]BF_4, as a new solvent, were described by Xiao and group. The ECL intensity of $Ru(bpy)_3^{2+}$ can be significantly increased in [Bmim]BF_4, especially existence of the TPA which can be inhibited by naphthol, and it might be used as sensitive method for the determination of naphthol with very low concentration and extends the analytical application of ECL for both the water-soluble and water-insoluble analytes [16]. ECL technique is a practical tool also for the study of protein folding, structure, and quantification. A highly sensitive approach for the detection of protein was proposed, employing bovine serum albumin (BSA) and casein. BSA and casein were found to be able to significantly quench the ECL of $Ru(bpy)_3^{2+}$/TPA system.

Inhibition mechanism was based on the formation of protein $Ru(bpy)_3^{2+}$ super molecule, which would prevent $Ru(bpy)_3^{2+}$ from reaching the working electrode surface to induce ECL quenching [17].

Quenching effect of ferrocene (Fc) is proved very practical for the development of ECL sensors. A reagentless signal-on ECL biosensor for DNA hybridization detection was developed (see Fig. 6.2). In the development of this biosensor, the quenching effect of Fc on intrinsic cathodic ECL at thin oxide-covered glassy carbon (C/C_xO_{1-x}) electrodes was employed. The main advantage of the present sensor lies in the fact that ECL is generated from the electrode itself and no luminophore or luminophore-labeled DNA probe is needed. The detection limit was ca. 5.0 pM (S/N = 3) [18].

Chen also employed ferrocene quenching of Ru-SNPs ECL for the fabrication of a novel and reusable signal-on aptasensor for adenosine. The biosensor has several advantages of good selectivity, high sensitivity, reproducibility, stability, and reusability [19]. Wang introduced an efficient solid-state ECL biosensing switch based on special ferrocene-labeled molecular beacon (Fc-MB) for DNA hybridization detection [20] and T4 DNA ligase detection [21]. The system consisted of two main parts: an ECL substrate and an ECL intensity switch. The ECL substrate was made by modifying the complex of Au nanoparticle and ruthenium (II) tris-(bipyridine) ($Ru(bpy)_3^{2+}$-AuNPs) onto Au electrode. And an ECL intensity switch is the molecular beacon labeled by ferrocene. The conformation change information of the Fc-MB before or after the ligation was a reflection of the ECL signal changes of the whole modified electrode. Thus, the ECL signal changes of the modified electrode can be detected once the nucleic acids were ligated. By this way, the ligation process can be monitored simultaneously and accurately. This method provides a new insight to investigate a wide variety of nucleic acids ligation processes and interactions between protein (enzyme) and nucleic acid.

Another ECL quenching through capture of Fc-labeled ligand-bound aptamer molecular beacon (MB) (see Fig. 6.3 for schematic illustration of the stepwise fabrication of aptasensor) was employed for an ultrasensitive protein-detection

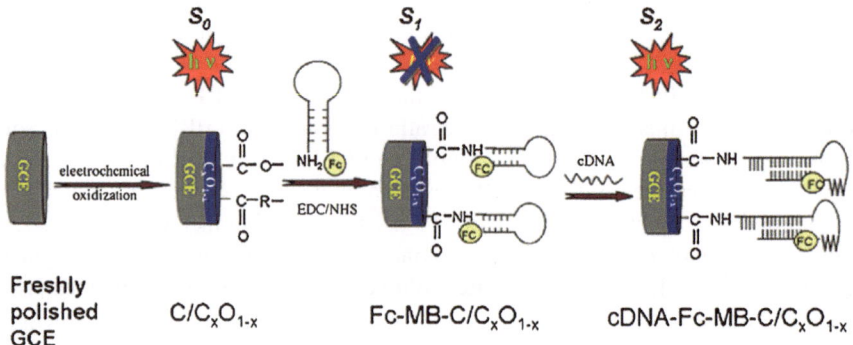

Fig. 6.2 Schematic diagram of the reagentless Fc-MB DNA biosensor. Reprinted with permission from Ref. [18]. Copyright 2010 Elsevier BV

protocol. The strategy consists of two main parts: a solid-state ECL sensing platform and an ECL switch. The sensing platform was constructed by modifying the composite of $Ru(bpy)_3^{2+}$ and platinum nanoparticles (Ru–PtNPs), followed by immobilization of capture DNA (CaDNA). MB worked as the ECL intensity switch. The present ECL quenching strategy for protein detection based on capture of ligand-bound MB directly in the solution offers much higher sensitivity with a detection limit of 1.7 pM. Thus, this procedure has great promise for ultrasensitive detection of disease markers directly in the solution [22].

A solid-state ECL protein biosensor developed by applying substitution strategy fulfills the successful detection of thrombin by using indirect labeling ECL method. The presence of thrombin induced the structure transformation of Fc-MB, consequently quenching the ECL from $Ru(bpy)_3^{2+}$-modified electrode. Figure 6.4 displays the scheme of the ECL biosensor based on the aptamer substitute strategy and ECL quenching effect. This biosensor can be developed to detect other proteins and biomolecules by using different probes [23].

Cao's group in 2011 designed an ECL biosensor based on the construction of triplex DNA for the detection of adenosine which employs an aptamer as a molecular recognition element and quenches ECL of $Ru(bpy)_3^{2+}$ by ferrocene monocarboxylic acid (FcA) (Fig. 6.5). In the presence of adenosine, the aptamer sequence (Ru-DNA-1) more likely forms the aptamer–adenosine complex with hairpin configuration and the switch of the DNA-1 occurs in conjunction with the

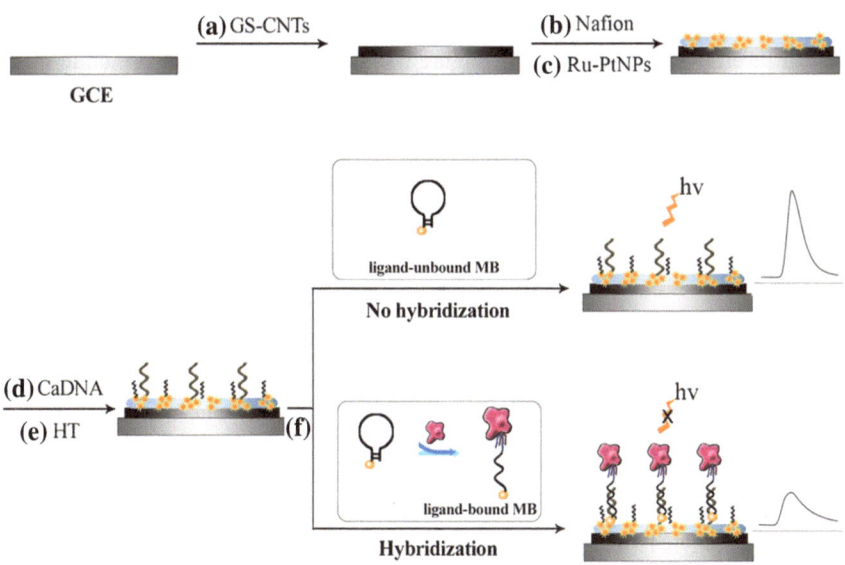

Fig. 6.3 The schematic illustration of the stepwise fabrication of aptasensor. **a** Loading of GS-CNTs, **b** formation of Nafion film, **c** immobilization of Ru–PtNPs, **d** adsorption of CaDNA, **e** blocking with HT, and **f** incubated in the ligand-bound MB solution. Reprinted with permission from Ref. [22]. Copyright 2011 Elsevier BV

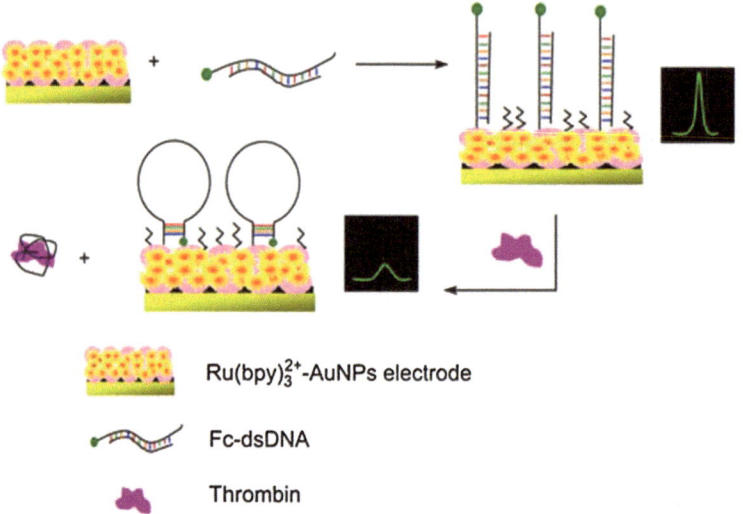

Ru(bpy)$_3^{2+}$-AuNPs electrode

Fc-dsDNA

Thrombin

Fig. 6.4 The scheme of the ECL biosensor based on the aptamer substitute strategy and ECL quenching effect. Reprinted with permission from Ref. [23]. Copyright 2011 Springer-Verlag

generation of a strong ECL signal owing to the dissociation of a quenching probe. The biosensor incorporates relative sensitive adenosine detection with a low detection limit of 2.7×10^{-10} mol L^{-1} [24].

A novel signal-on junction-probe-based ECL aptamer biosensor for the detection of thrombin was constructed. Figure 6.6 shows experimental principle of the junction-probe ECL aptamer biosensor for the detection of thrombin. Compared with some other ECL aptamer sensors, the proposed ECL aptamer biosensor has the advantages of higher sensitivity, lower background current, better selectivity and reusability, fabrication easiness, and operation convenience. These features make it a promising alternative to conventional thrombin detection methods and this method is expected to be involved in biochip fabrication [25].

Hanson's group described an electron paramagnetic resonance (EPR) study of the selective excitation and quenching mechanisms of Ru(bpy)$_3^{2+}$ CL and ECL. The results are complementary to the excited-state quenching mechanism postulated by Richter's group, who showed that quenching persisted when the benzoquinone derivatives of the tested phenols were initially present in the ECL system [26]. A FI procedure with inhibited ECL detection for determination of adrenaline has been illustrated. Adrenaline is found to inhibit strongly the ECL from the Ru(bpy)$_3^{2+}$/TPA system when a working Pt electrode is maintained at 1.05 V (versus Ag/AgCl) in pH 8.0 phosphate buffer. The method is proved to exhibit a good reproducibility, sensitivity, and stability, with a detection limit (S/N = 3) of 7.0×10^{-9} mol L^{-1}. The proposed strategy is based on energy transfer between freshly electrogenerated Ru(bpy)$_3^{2+}$ and the oxidation products such as adrenochrome and adrenaline quinone [27]. Guo et al. developed an ECL inhibition

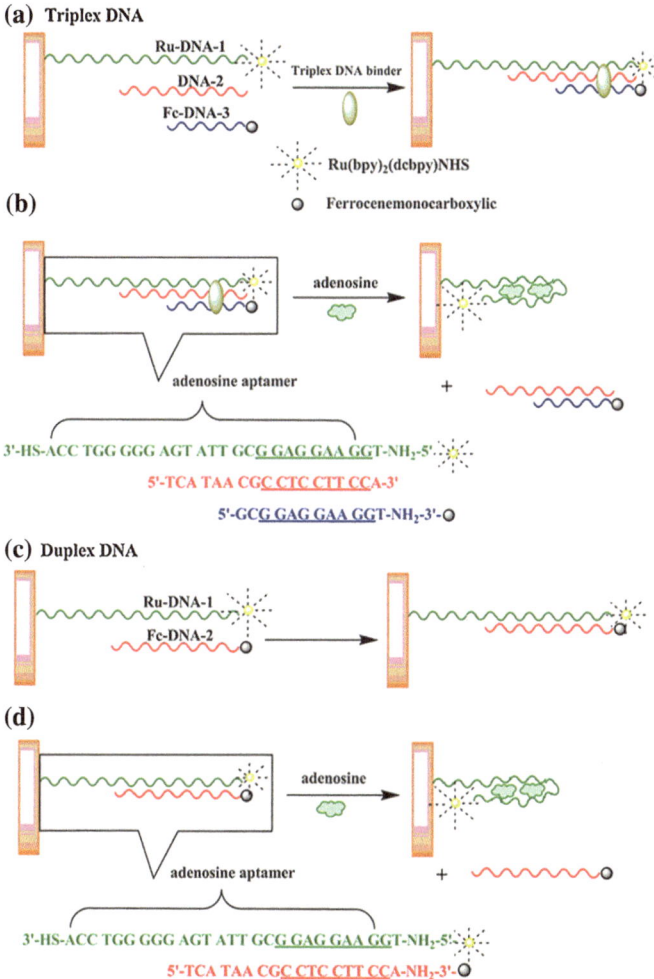

Fig. 6.5 Schematics of the ECL-triplex DNA biosensor (**a**) and the application in detection of adenosine (**b**), duplex DNA biosensor (**c**), and the application in detection of adenosine (**d**). Reprinted with permission from Ref. [24]. Copyright 2011 Elsevier BV

method for quantitative determination of malachite green (MG) residues in fish. This method is integrated with MISPE and was found that MG could strongly inhibit the ECL signal of luminol (mechanism shown in Fig. 6.7). By carrying out the oxidation reaction with 2, 3-dichloro-5, 6-dicyano-1, 4-benzoquinone (DDQ), which converts leucomalachite green (LMG) into MG, selective extraction and purification of MG were achieved with LOD 6 ppt [28].

Ruthenium ECL quenching has also been used for the detection of NO. This method was based on the FI equipped with a homemade gas pressure pump with low noise and a flow-through ECL cell. NO quenched the ECL emission of

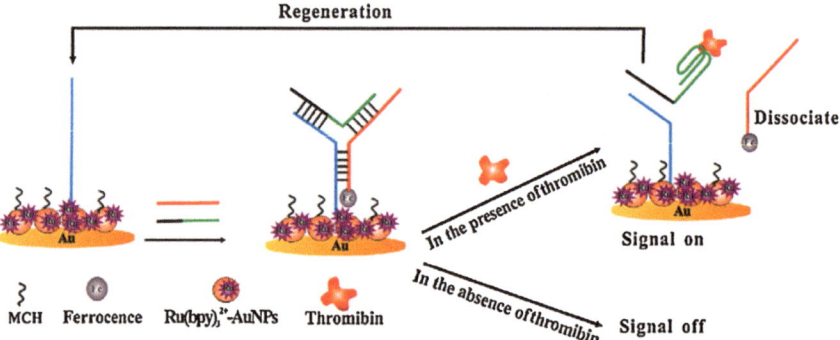

Fig. 6.6 Experimental principle of the junction-probe ECL aptamer biosensor for the detection of thrombin. Reprinted with permission from Ref. [25]. Copyright 2010 Elsevier BV

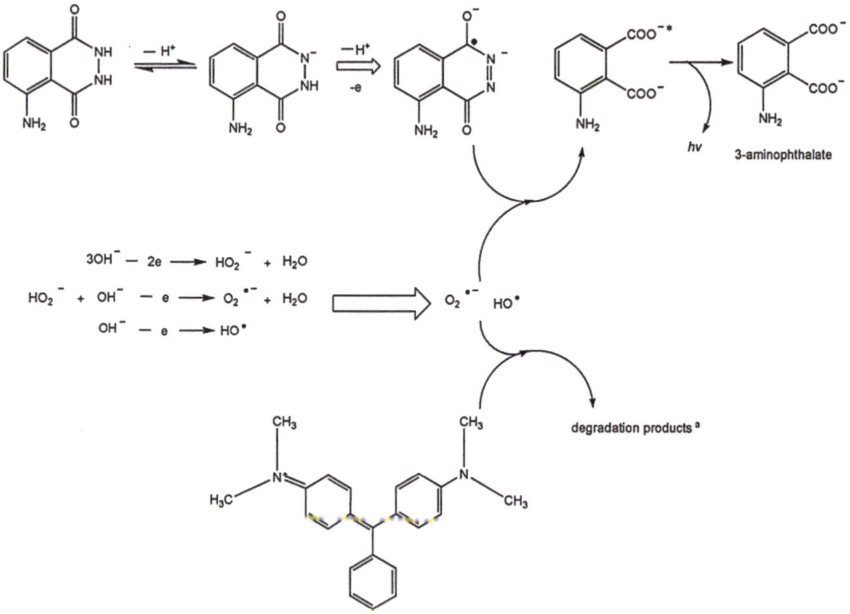

$$3OH^- - 2e \longrightarrow HO_2^- + H_2O$$

$$HO_2^- + OH^- - e \longrightarrow O_2^{\bullet -} + H_2O$$

$$OH^- - e \longrightarrow HO^{\bullet}$$

3-aminophthalate

$O_2^{\bullet -}$ $HO^{\bullet}$

degradation products ^{a}

Fig. 6.7 Mechanism for the ECL inhibition effect of MG on luminol. Reprinted with permission from Ref. [28]. Copyright 2011 American Chemical Society

$Ru(bpy)_3^{2+}$ at the oxidation potential of $Ru(bpy)_3^{2+}$ and set the detection limit of NO at 0.1 % of concentration of $Ru(bpy)_3^{2+}$ [29]. Inhibition phenomena of some catechol derivatives to the ECL of luminol were studied by Sun, and an FI-ECL inhibition method has been developed for the determination of catechol, 3,4-dihydroxybenzoic acid, and chlorogenic acid. Inhibition of ECL signal occurs on reaction of catechol derivatives with the freshly electrogenerated oxygen species on the electrode surface. The method provides more sensitivity and wider dynamic range than conventional spectrophotometric method or chemiluminescent method and has been successfully applied to determine chlorogenic acid in cigarettes [30].

Total amount of tetracycline (TC) residues was determined employing an ECL inhibition method by Guo [31]. An inhibited FI-ECL method for TC determination is also reported where energy transfer between electrogenerated benzoquinone derivatives and $Ru(bpy)_3^{2+*}$ on the electrode surface might be the cause of this inhibition. Figure 6.8 shows the electrochemiluminescence reactions of $Ru(bpy)_3^{2+}/TPrA/TCs$ [32]. QDs have also been employed as a tool in the detection of some analytes based on quenching mechanism. A new method with high sensitivity for determination of gossypol was developed based on the quenching effect of gossypol on the CdTe QD ECL emission which provided some new ideas for quick and sensitive detection [33].

ECL of CdTe/CdS QDs with tripropylamine as co-reactant based on quenching mechanism in aqueous solution was studied at a lower potential, and its application for highly sensitive and selective detection of Cu^{2+} was investigated [34]. An efficient ECL quenching achieved by functionalized CdTe QDs through ECL energy scavenging is also reported. This strategy is based on ultrasensitive antigen detection. The activated CdTe QDs have large absorption cross-section, act like

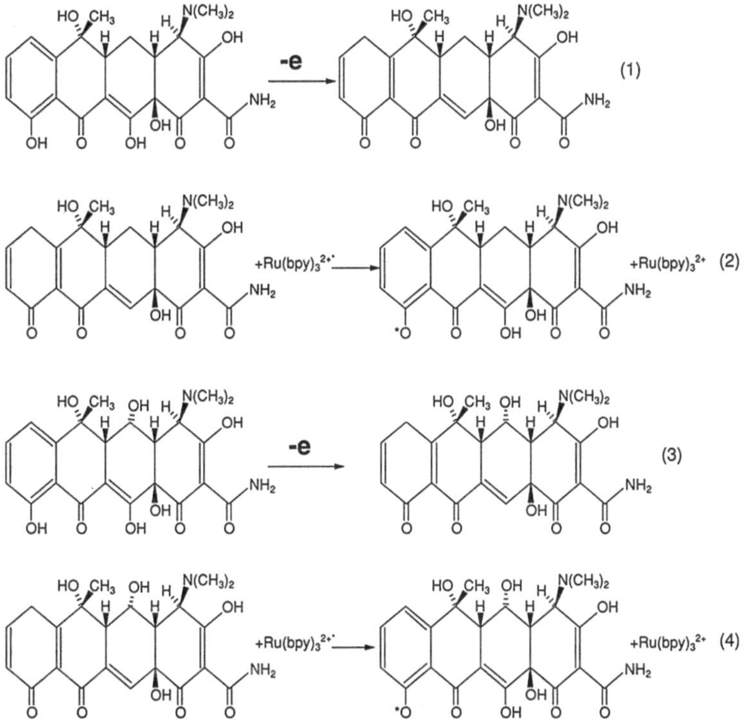

Fig. 6.8 Electrochemiluminescence reactions of $Ru(bpy)_3^{2+}/TPrA/TCs$

blackbodies, and effectively scavenge ECL energy. These blackbody-like CdTe QDs could have potential to find ECL quenchers for bioanalysis [35].

Another paper reported quenching of ECL emission from CdS:Mn NCs film by CdTe QDs. This strategy follows an energy scavenging process which could be amplified by the incorporation of a large number of QDs into the silica matrix. Figure 6.9 shows preparation of CdTe/SiO$_2$/pDNA composites and schematic diagram showing the principle of an aptamer-based assay for thrombin using CdTe/SiO$_2$ NPs as ECL quenching labels. As energy scavenging was a long-distance interaction, amplification of ECL quenching occurs. This long-distance ECL energy quenching shows a great potential in single-molecule detection [36].

A new reagentless ECL biosensor for hydroquinone (HQ) by co-immobilizing the (meso-2,3-dimercaptosuccinic acid) DMSA–CdTe QDs and horseradish per-oxidase (HRP) on a glassy carbon electrode, based on its quenching effect to the ECL emission, was demonstrated. The ECL quenching was due to the

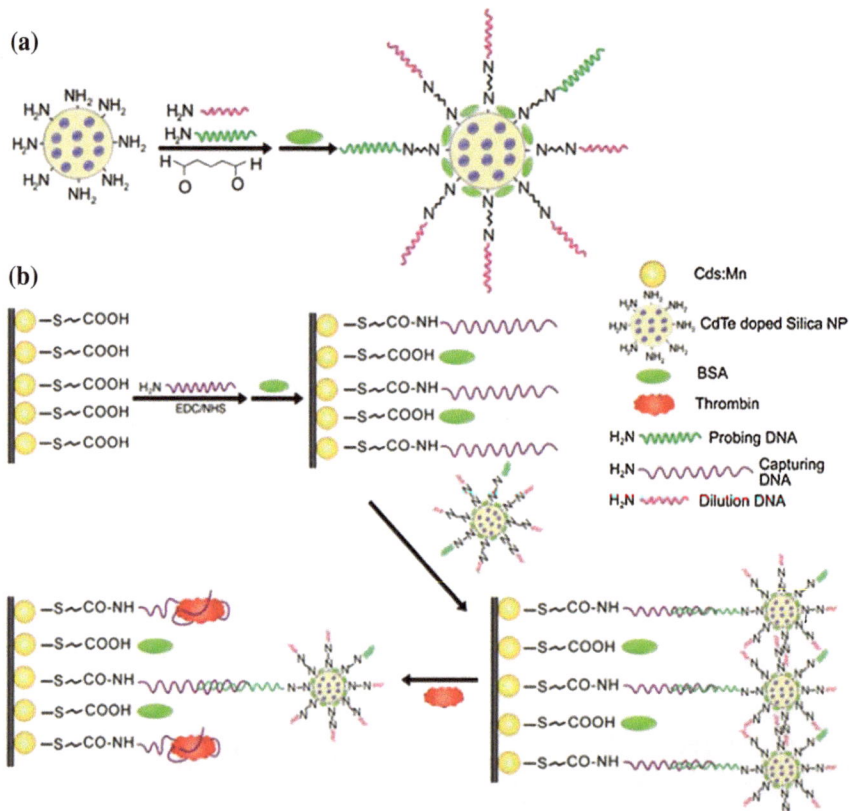

Fig. 6.9 a Preparation of CdTe/SiO$_2$/pDNA composites; **b** schematic diagram showing the principle of an aptamer-based assay for thrombin using CdTe/SiO$_2$ NPs as ECL quenching labels. Reprinted with permission from Ref. [36]. Copyright 2011 The Royal Society of Chemistry

consumption of H_2O_2 in the enzymatic cycle. The biosensor opens new avenues to the design of ECL emitters with low applied potential and would be favorable to bioanalysis for a wide range of analytes [37]. Water-soluble and positively charged 2-(dimethylamino)ethanethiol (DAET)-protected core–shell CdSe/ZnS QDs can effectively be incorporated within Nafion films by electrostatic interactions between the negatively charged sulfonic groups of Nafion and the positively charged DAET-protected core–shell CdSe/ZnS QDs. The quenching of the ECL emission derived from the Nafion/QDs composite film using hydrogen peroxide as a co-reactant scaled linearly with the concentration of glutathione (GSH), and the detection limit of concentrations up to 68 ($\pm$0.7) mM is obtained. This method of detection opens up the possibility to use Nafion/QDs composite films for various electroanalytical applications [38]. The quenching effect of DNA on the ECL of luminol was reported. A composite of CNTs and Au nanoparticles (AuNPs) was synthesized in order to achieve a novel and valuable label-free approach for DNA sensing. The ECL intensity remarkably decreases on absorption of more than 1.0×10^{-12} mol of DNA. The proposed mechanism for the quenching effect is ascribed to the interaction between luminol and DNA and the elimination of reactive oxygen species (ROSs) by DNA. The dsDNA could efficiently quench the ECL of luminol in the pH range of 6.5–11. The DNA sensor furnishes exceptional stability and reproducibility and could potentially serve as a powerful tool for the label-free investigation of dsDNA and might encourage the further development and applications in clinical practice, medicine, and basic research [39].

DA is determined in pharmaceutical injections on the basis of its inhibition of the ECL of luminol. A simple and sensitive ECL-FI method for DA detection has been developed linearly with DA concentration in the range of 5.0×10^{-8} – 1.0×10^{-5} mol L^{-1} with a detection limit of 30 nmol L^{-1} [40]. An indirect ECL inhibition method integrated with CE was developed for the determination of DA and epinephrine using a 10 mM phosphate –0.5 mM TPA (pH 9.0) solution as the running buffer and 50 mM Ru(bpy)$_3^{2+}$ in 50 mM phosphate electrolyte (pH 8.5) as luminescence reagent. Based on the quenching effect of the Ru(bpy)$_3^{2+}$/TPA system, this method could be used for separation and detection of analytes at trace level [41]. A nanoliter-sized flow cell is developed for construction of an FIA system with ECL detection based on sensitive quenching response of DA. This system falls within the linear range from 10 pM to 4 nM and realizes a detection limit as low as 3.6 pM. The present approach is superior to corresponding analysis methods in sensitivity, linear range, and reproducibility and is practically used to determine the neurotransmitters in cerebrospinal fluid (CSF) with DA as the index giving satisfactory results [42].

Another method based on FI-ECL inhibition method for determination of neurotransmitters, noradrenaline, and DA has been developed for Ru(bpy)$_3^{2+}$/TPA and for Ru(phen)$_3^{2+}$/TPA systems. The proposed inhibition mechanism shows the quenching effect of o-benzoquinone derivative generated from the direct electro-oxidation reactions of noradrenaline and DA. This method has superiority of sensitivity over several detection methods, such as spectrometry, electro-chemical, CE, and HPLC methods, and has potential for sensitive and trace analyses of commercial pharmaceutical injection samples with satisfied results [43]. Study of some other literature revealed 2-(dibutylamino)ethanol (DBAE) to be a more efficient co-reactant as compared to the tripropylamine for ECL of $Ru(bpy)_3^{2+}$. DA was found to be able to inhibit the ECL of $Ru(bpy)_3^{2+}$/DBAE system strongly with the detection limit of 4.0×10^{-11} M, based on which an ultrasensitive approach for DA detection could be established. Thus, $Ru(bpy)_3^{2+}$/DBAE-DA ECL system should be of great potential for its application in analytical chemistry [44].

An FI-ECL method based on an inhibition effect on the $Ru(bpy)_3^{2+}$/TPA has been demonstrated for the simple and convenient determination of gallic acid with a determination limit of 9.0×10^{-9} mL^{-1}. Most probably, the interaction between electrogenerated $Ru(bpy)_3^{2+*}$ and o-benzoquinone derivative generated from the electrode oxidation of GA at the electrode surface becomes the reason for inhibition [45]. Indirect determination of glucose based on the linear relationship between concentration of H_2O_2 and the decrease in ECL intensity in a $Ru(bpy)_3^{2+}$/TPA system has been introduced. Glucose solutions are run through a glucose oxidase (GOD)-immobilized solgel column and an ECL system of $Ru(bpy)_3^{2+}$ and TPA. Glucose is determined with a detection limit of 1.0 μM in a linear dynamic range of 1.0–200 μM. This technique has so many advantages such as high sensitivity, lower detection limit, and capable of eliminating or minimizing interference to an insignificant level; thus, the proposed ECL system is promising in the analysis of other important clinical or biomedical analytes which produce hydrogen peroxide as its enzymatic product, namely cholesterol, galactose, choline, methanol, uric acid. [46].

Tannic acid is found to strongly inhibit the ECL of luminol, on the basis of which another simple, rapid, and sensitive FI method for the determination of tannic acid has been developed [47]. Biocatalytic precipitation (BCP) was firstly integrated with ECL to explore its insulating effect on ECL, intending to discover an alternative ECL highly efficient quenching route. Inhibition of the reaction between the co-reactant and luminophore takes place due to an insulating layer formation via BCP onto the electrode surface, inhibiting the reaction between the co-reactant and luminophore and thereby impairing the ECL emission dramatically. The proposed method was applied in HRP-based system with a linear range from 1.0×10^{-10} to 1.0×10^{-6} M for H2O2 determination, and a detection limit of 4×10^{-11} M (S/N = 3) was realized [48].

Efficient ECL quenching of $Ru(bpy)_3^{2+}$/TPA system has been reported using a novel quencher caffeic acid (CA). Quenching mechanism involves in the energy transfer from the excited-state $Ru(bpy)_3^{2+}$ to the electrochemical oxidation of CA, which is produced at the electrode surface. Additionally, this strategy may find its use in environmental analyses of CA and its derivatives as well as in biotechnology and pharmaceutical applications [49]. In recent reports, uric acid (UA) determination based on the quenching of ECL was also illustrated. This method has the advantage of detecting UA at LOP and leads to the development of a new method for the detection of UA in biological samples [50].

References

1. Bard AJ (2004) Electrogenerated chemiluminescence. Taylor & Francis, London
2. Kuhn H (1970) Classical aspects of energy transfer in molecular systems. J Chem Phy 53(1):101–108. doi:10.1063/1.1673749
3. Chance RR, Prock A, Silbey R (1975) Comments on the classical theory of energy transfer. J Chem Phy 62(6):2245–2254. doi:10.1063/1.430748
4. Chandross EA, Visco RE (1968) Preannihilation electrochemiluminescence and the heterogeneous electrochemical formation of excited states. J Phy Chem 72(1):378–379
5. Greenway GM, Knight AW, Knight PJ (1995) Electrogenerated chemiluminescent determination of codeine and related alkaloids and pharmaceuticals with tris(2,2[prime or minute]-bipyridine)ruthenium(II). Analyst 120(10):2549–2552
6. Zheng H, Zu Y (2005) Highly efficient quenching of coreactant electrogenerated chemiluminescence by phenolic compounds. J phy chem B 109(33):16047–16051. doi:10.1021/jp052843o
7. Zheng H, Zu Y (2005) Emission of tris(2,2'-bipyridine)ruthenium(II) by coreactant electrogenerated chemiluminescence: from O2-insensitive to highly O2-sensitive. J physl chem B 109(24):12049–12053. doi:10.1021/jp050350d
8. Pizzocaro C, Ml Bolte, Hoffman MZ (1992) $Cr(bpy)_3^{3+}$ -sensitized photo-oxidation of phenol in aqueous solution. J Photochem Photobio A 68(1):115–119
9. Brune SN, Bobbitt DR (1992) Role of electron-donating/withdrawing character, pH, and stoichiometry on the chemiluminescent reaction of tris(2,2'-bipyridyl)ruthenium(III) with amino acids. Anal Chem 64(2):166–170
10. McCall J, Alexander C, Richter MM (1999) Quenching of electrogenerated chemiluminescence by phenols, hydroquinones, catechols, and benzoquinones. Anal Chem 71(13):2523–2527
11. McCall J, Richter MM (2000) Phenol substituent effects on electrogenerated chemiluminescence quenching. Analyst 125(3):545–548
12. Chi Y, Dong Y, Chen G (2007) Inhibited $Ru(bpy)_3^{2+}$ electrochemiluminescence related to electrochemical oxidation activity of inhibitors. Anal Chem 79(12):4521–4528. doi:10.1021/ac0702443
13. Qiu B, Xue L, Wu Y, Lin Z, Guo L, Chen G (2011) Mechanism study on inorganic oxidants induced inhibition of $Ru(bpy)_3^{2+}$) electrochemiluminescence and its application for sensitive determination of some inorganic oxidants. Talanta 85(1):339–344. doi:10.1016/j.talanta.2011.03.063
14. Guo L, Xue L, Qiu B, Lin Z, Kim D, Chen G (2010) Mechanism study on inhibited $Ru(bpy)_3^{2+}$ electrochemiluminescence between coreactants. Phys Chem Chem Phy 12(39):12826–12832. doi:10.1039/c004277c
15. Li X, Sun L, Ding T (2011) Multiplexed sensing of mercury(II) and silver(I) ions: a new class of DNA electrochemiluminescent-molecular logic gates. Biosens Bioelectron 26(8):3570–3576. doi:10.1016/j.bios.2011.02.003
16. Yu X, Dai J, Yang L, Xiao D (2010) 1-Butyl-3-methylimidazolium based ionic liquid as solvent for determination of hydrophobic naphthol with the electrogenerated chemiluminescence of tris(2,2'-bipyridine) ruthenium(II). Analyst 135(3):630–635. doi:10.1039/b916435a
17. Qiu B, Jiang X, Guo L, Lin Z, Cai Z, Chen G (2011) A highly sensitive method for detection of protein based on inhibition of $Ru(bpy)_3^{2+}$/TPrA electrochemiluminescent system. Electrochim Acta 56(20):6962–6965. doi:10.1016/j.electacta.2011.06.016
18. Wu A-H, Sun JJ, Zheng RJ, Yang HH, Chen GN (2010) A reagentless DNA biosensor based on cathodic electrochemiluminescence at a $C/C_{(x)}O_{(1-x)}$ electrode. Talanta 81(3):934–940. doi:10.1016/j.talanta.2010.01.040

19. Chen L, Cai Q, Luo F, Chen X, Zhu X, Qiu B, Lin Z, Chen G (2010) A sensitive aptasensor for adenosine based on the quenching of $Ru(bpy)_3^{2+}$-doped silica nanoparticle ECL by ferrocene. Chem Comm 46(41):7751–7753. doi:10.1039/c0cc03225e

20. Wang X, He P, Fang Y (2010) A solid-state electrochemiluminescence biosensing switch for detection of DNA hybridization based on ferrocene-labeled molecular beacon. J Lumin 130(8):1481–1484. doi:10.1016/j.jlumin.2010.03.016

21. Wang X, Dong P, Yun W, Xu Y, He P, Fang Y (2010) Detection of T4 DNA ligase using a solid-state electrochemiluminescence biosensing switch based on ferrocene-labeled molecular beacon. Talanta 80(5):1643–1647. doi:10.1016/j.talanta.2009.09.060

22. Liao Y, Yuan R, Chai Y, Mao L, Zhuo Y, Yuan Y, Bai L, Yuan S (2011) Electrochemiluminescence quenching via capture of ferrocene-labeled ligand-bound aptamer molecular beacon for ultrasensitive detection of thrombin. Sens Actuators B 158(1):393–399. doi:10.1016/j.snb.2011.06.045

23. Xu Y, Dong P, Zhang X, He P, Fang Y (2011) Solid-state electrochemiluminescence protein biosensor with aptamer substitution strategy. Sci China-Chem 54(7):1109–1115. doi:10.1007/s11426-011-4278-y

24. Ye S, Li H, Cao W (2011) Electrogenerated chemiluminescence detection of adenosine based on triplex DNA biosensor. Biosens Bioelectron 26(5):2215–2220. doi:10.1016/j.bios.2010.09.037

25. Zhang J, Chen P, Wu X, Chen J, Xu L, Chen G, Fu F (2011) A signal-on electrochemiluminescence aptamer biosensor for the detection of ultratrace thrombin based on junction-probe. Biosens Bioelectron 26(5):2645–2650. doi:10.1016/j.bios.2010.11.028

26. Hindson CM, Hanson GR, Francis PS, Adcock JL, Barnett NW (2011) Any old radical won't do: an EPR study of the selective excitation and quenching mechanisms of $Ru(bipy)_3^{2+}$ chemiluminescence and electrochemiluminescence. Chem Eur J 17(29):8018–8022. doi:10.1002/chem.201100877

27. Li F, Cui H, Lin XQ (2002) Determination of adrenaline by using inhibited $Ru(bpy)_3^{2+}$ electrochemiluminescence. Anal Chim Acta 471(2):187–194. doi:10.1016/s0003-2670(02)00930-3

28. Guo Z, Gai P, Hao T, Duan J, Wang S (2011) Determination of malachite green residues in fish using a highly sensitive electrochemiluminescence method combined with molecularly imprinted solid phase extraction. J Agr Food Chem 59(10):5257–5262. doi:10.1021/jf2008502

29. Chen J, Miyake M, Chi Y, Nishiumi T, Aoki K (2007) Determination of nitric oxide by quenching electro-chemiluminescence of tris(2,2′-bipyridyl)ruthenium in flow injection analysis. Electroanalysis 19(2–3):181–184. doi:10.1002/elan.200603689

30. Sun YG, Cui H, Li YH, Lin XQ (2000) Determination of some catechol derivatives by a flow injection electrochemiluminescent inhibition method. Talanta 53(3):661–666. doi:10.1016/s0039-9140(00)00550-6

31. Guo Z, Gai P (2011) Development of an ultrasensitive electrochemiluminescence inhibition method for the determination of tetracyclines. Anal Chim Acta 688(2):197–202. doi:10.1016/j.aca.2010.12.043

32. Pang YQ, Cui H, Zheng HS, Wan GH, Liu LJ, Yu XF (2005) Flow injection analysis of tetracyclines using inhibited $Ru(bPY)_3^{2+}$/tripropylamine electrochemiluminescence system. Luminescence 20(1):8–15. doi:10.1002/bio.793

33. Hua L, Zhou J, Han H (2010) Direct electrochemiluminescence of CdTe quantum dots based on room temperature ionic liquid film and high sensitivity sensing of gossypol. Electrochim Acta 55(3):1265–1271. doi:10.1016/j.electacta.2009.10.038

34. Mei YL, Wang HS, Li YF, Pan ZY, Jia WL (2010) Electrochemiluminescence of CdTe/CdS quantum dots with tripropylamine as coreactant in aqueous solution at a lower potential and its application for highly sensitive and selective detection of Cu^{2+}. Electroanalysis 22(2):155–160. doi:10.1002/elan.200904685

35. Shan Y, Xu J–J, Chen H-Y (2010) Electrochemiluminescence quenching by CdTe quantum dots through energy scavenging for ultrasensitive detection of antigen. Chem Comm 46(28):5079–5081. doi:10.1039/c0cc00837k

36. Shan Y, Xu J–J, Chen H-Y (2011) Enhanced electrochemiluminescence quenching of CdS:Mn nanocrystals by CdTe QDs-doped silica nanoparticles for ultrasensitive detection of thrombin. Nanoscale 3(7):2916–2923. doi:10.1039/c1nr10175g

37. Liu X, Cheng L, Lei J, Liu H, Ju H (2010) Formation of surface traps on quantum dots by bidentate chelation and their application in low-potential electrochemiluminescent biosensing. Chem Eur J 16(35):10764–10770. doi:10.1002/chem.201001738

38. Dennany L, Gerlach M, O'Carroll S, Keyes TE, Forster RJ, Bertoncello P (2011) Electrochemiluminescence (ECL) sensing properties of water soluble core-shell CdSe/ZnS quantum dots/Nafion composite films. J Mat Chem 21(36):13984–13990. doi:10.1039/c1jm12183a

39. Chu H–H, Yan J-L, Tu Y-F (2010) Study on a luminol-based electrochemiluminescent sensor for label-free DNA sensing. Sensors 10(10):9481–9492. doi:10.3390/s101009481

40. Zhu LD, Li YX, Zhu GY (2002) Flow injection determination of dopamine based on inhibited electrochemiluminescence of luminol. Anal Lett 35(15):2527–2537. doi:10.1081/al-120016542

41. Kang JZ, Yin XB, Yang XR, Wang EK (2005) Electrochemiluminescence quenching as an indirect method for detection of dopamine and epinephrine with capillary electrophoresis. Electrophoresis 26(9):1732–1736. doi:10.1002/elps.200410247

42. Zhao J, Chen M, Yu C, Tu Y (2011) Development and application of an electrochemiluminescent flow-injection cell based on CdTe quantum dots modified electrode for high sensitive determination of dopamine. Analyst 136(19):4070–4074. doi:10.1039/c1an15458c

43. Li F, Pang YQ, Lin XQ, Cui H (2003) Determination of noradrenaline and dopamine in pharmaceutical injection samples by inhibition flow injection electrochemiluminescence of ruthenium complexes. Talanta 59(3):627–636. doi:10.1016/s0039-9140(02)00576-3

44. Xue L, Guo L, Qiu B, Lin Z, Chen G (2009) Mechanism for inhibition of/DBAE electrochemiluminescence system by dopamine. Electrochem Commun 11(8):1579–1582

45. Lin XQ, Li F, Pang YQ, Cui H (2004) Flow injection analysis of gallic acid with inhibited electrochemiluminescence detection. Anal Bioanal Chem 378(8):2028–2033. doi:10.1007/s00216-004-2519-z

46. Wang CY, Huang HJ (2003) Flow injection analysis of glucose based on its inhibition of electrochemiluminescence in a Ru(bpy)$_3^{2+}$-tripropylamine system. Anal Chim Acta 498(1–2):61–68. doi:10.1016/j.aca.2003.08.064

47. Sun YG, Cui H, Li YH, Zhao HZ, Lin XQ (2000) Flow injection analysis of tannic acid with inhibited electrochemiluminescent detection. Anal Lett 33(11):2281–2291. doi:10.1080/00032710008543189

48. Wang J, Zhao WW, Tian CY, Xu JJ, Chen HY (2012) Highly efficient quenching of electrochemiluminescence from CdS nanocrystal film based on biocatalytic deposition. Talanta 89:422–426

49. Zhu Y, Zhao B, Li L, Chen W, Tang W, Zhao G (2010) Quenching of the electrochemiluminescence of tris(2,2′-bipyridine)ruthenium(ii) by caffeic acid. Anal Lett 43(13):2105–2113. doi:10.1080/00032711003698804

50. Chen Z, Zu Y (2008) Selective detection of uric acid in the presence of ascorbic acid based on electrochemiluminescence quenching. J Electroanal Chem 612(1):151–155. doi:10.1016/j.jelechem.2007.09.018

Chapter 7
Applications of Electrochemiluminescence

Abstract The development of electrochemiluminescence (ECL) applications is a growing field, having the potential advantages of ECL over conventional chemi-luminescence. ECL has found various applications in immunoassays, DNA probe assays, and aptasensors by employing ECL-active species as labels on biological molecules. The recent use of ECL to detect many chemically, biochemically, clinically, and environmentally important analytes is reviewed.

Keywords Clinical diagnostics · Biowarfare agent detection · Pharmaceutical analysis · Environmental analysis · Immunoassays · DNA biosensor · Aptasensor

For last few years, electrochemiluminescence (ECL) has been coupled with many techniques. Biochemical and analytical applications of ECL including medical analysis and clinical diagnostics, in biosensors, in trace analysis environmental assays such as food and water testing, biowarfare agent detection and as ECL markers have been gaining rapid significance for last decade. It is useful not only in biology and biochemistry, but also in polymer sciences and in membranes research. In addition to this, some other related topics are developing like ECL imaging, ECL in semiconductors, ECL in scanning microscopy at ultramicro-electrodes, ECL in material chemistry for light-emitting devices, and others. ECL and many other techniques coupled together provide rapid, efficient, and versatile methods of separation and detection of biochemicals using micro, nano, or femto volumes of sample and has become a very sensitive, powerful, and promising method [1].

7.1 Clinical Applications/Immunosensors

ECL, a highly sensitive technique, has a highly significant, most common and, arguably, the most important commercial application in diagnostic assays. More than 150 ECL different immunoassays [2] including those for thyroid diseases, tumor and cardiac markers, fertility therapies, and analytes relevant to infection diseases have commercial value and are currently available. Many efforts have been taken to improve sensitivity and extend applications of ECL immunoassays [3]. ECL intensity depends upon the concentration of emitter and analyte [4, 5], so can be used to analyze the concentration of both. Usually $Ru(bpy)_3^{2+}$ is used as an ECL label and the concentration of an analyte can be determined by measuring the emission of a label in the presence of a high and constant concentration of co-reactant while some assays rely on the detection of co-reactants in the presence of constant $Ru(bpy)_3^{2+}$ concentrations.

An ultrasensitive sandwich-type human C-reactive protein (CRP) immunoassay based on adding multiple $Ru(bpy)_3^{2+}$ to a single antibody was described encapsulating a hydrophobic $Ru(bpy)_3^{2+}$ compound in polystyrene microspheres/beads [6]. This technique leads the CRP detection limit as low as 0.01 mg mL^{-1}, which is lower than those obtained from most of the presently available automated high-sensitivity CRP assay systems. Surface proteins on live cells were also detected employing $Ru(bpy)_3^{2+}$ ECL [7]. In this method, carbon surface electrodes are used to build into the bottom of microwell plates to achieve ECL assay. The carbon surface plates bind suspension cells tightly enough to allow plates to be washed, avoiding the centrifugation steps typically used for washing suspension cells. Acetylcholinesterase has also been recently used as an alternative label in ECL immunoassays [8]. The thiocholine generated from acetylthiocholine using acetylcholinesterase was collected on gold electrode surface by gold–thiol binding. The ECL detection was performed using this thiocholine as a co-reactant in the presence of $Ru(bpy)_3^{2+}$. This method can greatly enhance the sensitivity since a large number of co-reactant molecules can be generated by the enzymatic reaction.

Novel, ultrasensitive, and renewable electrochemiluminescent immunosensors were constructed based on polymerization-assisted signal amplification for detection of cancer biomarkers [9], combining a newly designed trace tag and streptavidin-coated magnetic particles (SCMPs) for tumor markers [10], and $Ru(bpy)_3^{2+}$-encapsulated silica nanoparticle ($SiO_2@Ru$) as labels for biomarkers through a sandwiched immunoassay process [11]. A FI-ECL system for immunoassay of atrazine [12] and human immunogloblin G (hIgG) with a variety of templates [13–15] has been developed. Optical immunodetection of herbicide 2,4-D based on FI-ECL of luminal [16] and a miniature 96 sample ELISA-lab-on-a-chip (ELISA-LOC) for enzyme-linked immunodetection of staphylococcal enterotoxin B (SEB) [17] was fabricated. A microfluidic assay format was developed combined with a CCD sensor.

Another ultrasensitive multiplexed ECL immunoassay method was developed for the detection of tumor markers by combining functionalized graphene nanosheets and gold-coated magnetic Fe_3O_4 nanoparticles (GMPs) labeled alpha-fetoprotein (AFP) antibody [18]. An enzyme immunoassay for TNT (2,4,6-trinitrotoluene) detection was developed by Wilson in which enzyme-labeled antibodies bound to paramagnetic beads are concentrated on an electrode magnetically, and light emission is triggered [19, 20]. Another approach was developed for the determination of hemagglutinin which plays an important role in influenza virus infection hemagglutinin that was first immobilized on an Au electrode; an immunoliposome encapsulating a Ru complex was prepared to bind with hemagglutinin through competitive antigen–antibody reaction. After the immunoreaction, the immuno-liposome was destroyed by the addition of ethanol, and the ECL was measured from the Ru complex adsorbed onto Au electrode surfaces (Fig. 7.1). Hemagglutinin of influenza virus was determined in a concentration range from 3×10^{-13} to 4×10^{-11} g mL^{-1}. The ECL method with high sensitivity at the attomole level would be applicable for detecting trace amounts of various proteins containing the influenza virus [21].

Several immunoassays can now be performed with imaging-based ECL instruments manufactured by Meso Scale Discovery (Chap. 3, Fig. 3.7) [22, 23]. Approximately 150 immunoassays are currently available from this company. These include phosphoprotein and intracellular markers, cardiac markers, vascular markers and growth factors, fertility markers, Alzheimer's disease markers, hypoxia markers, toxicology, metabolic markers, bone markers, bioprocess assays, and cytokine and chemokine immunoassays. Further description and discussion on the particular immunoassays of interest may be found in Ref. [22].

ECL reaction of an acridinium ester, methyl-9-(p-formylphenyl) acridinium carboxylate fluorosulfonate (MFPA), on the surface of a platinum electrode has been used for immunoassay by the determination of human chorionic gonadotropin using MFPA as label [24]. A solution-phase quantitative ECL immunoassay was developed for one-step rapid measurement of antigen–antibody binding affinity. This solution-phase ECL immunoassay may be applied to detect other serum tumor markers or study protein–protein interactions [25]. Liu and group demonstrated a versatile immunosensor using a CdTe QDs coated silica nanosphere (Si/QD) as a label for ultrasensitive detection of a biomarker [26]. A water/oil microemulsion method was used to synthesize Ru(bpy)$_3$$^{2+}$-doped silica (RuSi) nanoparticles and

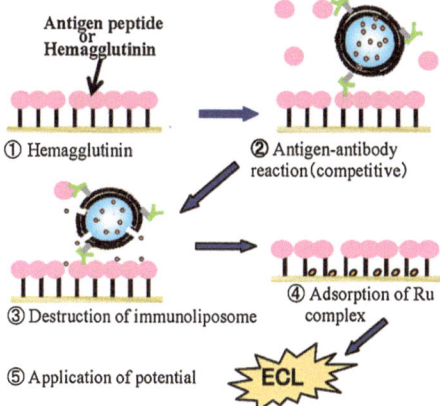

Fig. 7.1 Detection procedure: **a** Immobilization of hemagglutinin (or antigen peptide) on an Au electrode. **b** Binding of immunoliposome with hemagglutinin onto the Au electrode through competitive antigen–antibody reaction. **c** Destruction of immunoliposome by addition of ethanol. **d** Adsorption of Ru(II) complex by heating at 60 °C for 10 min. **e** ECL measurement on application of potential. Reprinted with permission from Ref. [21]. Copyright 2008 American Chemical Society

was immobilized on a glassy carbon electrode by using TPA as a co-reactant and used as tag for immunoassay and DNA detection [27].

7.2 Pharmaceutics/Criminalistic Expertize

Several research groups have made contributions to the development of ECL biosensors for applications in pharmaceutics. Many of these examples completely satisfy our requirements, while it seems that still there is need of development in pharmaceutical industry. Currently, various techniques in combination with ECL biosensing formats for the determination of drugs and small molecules have been developed. ECL combined with different systems has also been successfully applied to criminalistic expertize for sensitive determination of narcotics and medicinal preparations [28–33].

Tian and group reported a very sensitive method based on CE coupled with ECL detection with $Ru(bpy)_3^{2+}$ for simultaneous determination of azithromycin (AZI), acetylspiramycin (ACE), erythromycin (ERY), and josamycin (JOS). The four macrolides were well separated and detected within 6 min under the optimized conditions with LOD (S/N = 3) of AZI, ACE, ERY, and JOS as 1.2×10^{-9}, 7.1×10^{-9}, 3.9×10^{-8} and 9.5×10^{-8} mol L^{-1}, respectively [34]. A new separation and the quantification method for the sensitive determination of ephedrine and pseudoephedrine [35, 36] based on $Ru(bpy)_3^{2+}$ ECL coupled with CE has been established with detection limit of 4.5×10^{-9} g mL^{-1} and 4.5×10^{-8} mol L^{-1} for ephedrine and 5.2×10^{-8} mol L^{-1} (S/N = 3) for pseudoephedrine, respectively. The method was applied for the determination of ephedrine and pseudoephedrine in human urine, ephedrine in nasal drops, and the monitoring of pharmacokinetics for pseudoephedrine.

For direct analysis of ibandronate (IBAN), a simple, rapid, and sensitive CE method integrated with ECL detection has been reported. SPE using magnetic $Fe_3O_4@Al_2O_3$ nanoparticles as the extraction phase was employed to pretreat the urine sample before CE-ECL analysis. The linear range was 0.2–12.0-mM IBAN in human urine, and LOD was 0.06 mM [37]. An angiotensin-converting-enzyme (ACE) inhibitor, captopril, was determined employing a new FI-ECL procedure in a pharmaceutical preparation based on the direct CL oxidation of captopril by nascent Mn^{3+} with a detection limit of 8.0×10^{-8} mol L^{-1} original concentration [38]. Modifying a graphite electrode with vaseline and NiO, ranitidine, an H_2 blocker, showed a strong ECL for the weak ECL signal of electrooxidation of luminol. Based on this finding, a more sensitive ECL method for ranitidine was proposed which offered a 9×10^{-9} mol L^{-1} detection limit for ranitidine hydrochloride [39]. Michel's research group performed batch and continuous-flow measurements of codeine in pharmaceutical samples by ECL, using a miniaturized sensor combining transduction electrodes and a photodetector on the same silicon chip. In batch mode and FIA, a detection limit of 0.1 and 50 mM was found, respectively [40].

Furthermore, CE separation with end-column $Ru(bpy)_3^{2+}$ ECL detection for the quantitative determination of pentoxyverine (an antitussive) was performed. Gold nanoparticles were found to enhance the ECL intensity. The detection limits with and without nanogold were 6 nM and 0.6 mM, respectively. Successful separation of pentoxyverine, chlorpheniramine, and lidocaine was achieved. This method was also applied to monitor drug binding with HSA [41]. Determination of propranolol hydrochlorid, a β-adrenergic blocker, in human urine was performed under the optimal conditions with a linear range from 0.003 to 2 µg mL^{-1} ($r^2 = 0.9993$), and detection limit was 1.3 ng mL^{-1} (S/N = 3) [42]. Moreover, some β-blockers such as atenolol, metoprolol, and esmolol [43, 44] were simultaneously determined in commercial pharmaceutical, human body, and urine samples employing a novel ECL sensor combined with CE separation method. For a fine and improved separation, Poly-β-cyclodextrin (Poly-β-CD) was used as an additive in the running buffer [44].

Other literature presented a rapid method for the determination of Dioxopromethazine hydrochloride (DPZ) (an antihistamine drug) by CE–ECL using $Ru(bpy)_3^{2+}$ reagent. The proposed method was of high sensitivity, good selectivity, and reproducibility for DPZ analysis. This CE–ECL method was applied to analyze DPZ in real samples including tablets, rat serum, and human urine [45].

Fast analysis of ofloxacin and lidocaine (as bactericide and analgesic) is of clinic importance for understanding the patient's medical process. A sensitive method for the determination of lidocaine, ofloxacin [46], enrofloxacin (ENR), and ciprofloxacin (CIP) [47] by CE integrated with ECL detection has been developed based on porous etched joint and end-column $Ru(bpy)_3^{2+}$ ECL and successfully applied to determine ENR and CIP in milk with a solid-phase extraction procedure. The proposed method explores detection limits of lidocaine, ofloxacin, enrofloxacin, and ciprofloxacin as 3.0×10^{-8}, 5.0×10^{-7}, 10×10^{-9}, and 15×10^{-9} mol L^{-1}, respectively (S/N = 3). CE-ECL detection method has also been used to characterize disopyramide with a detection limit of 2.5×10^{-8} mol L^{-1} (S/N = 3) [48], and procaine hydrolysis as a probe for butyrylcholinesterase by in vitro procaine metabolism in plasma with butyrylcholinesterase acting as bioscavenger. Procaine and its metabolite N, N-diethylethanolamine were separated at 16 kV with the detection limits of 2.4×10^{-7} and 2.0×10^{-8} mol L^{-1} (S/N = 3), respectively [49].

Several antibiotics of clinical and pharmaceutical importance were also determined on the basis of ECL detection method. For instance, based on the enhancing effect of clarithromycin on ECL of $Ru(bpy)_3^{2+}$ after CE separation, a new analytical method for sensitive determination of clarithromycin in biological fluids was developed and a detection limit of 30 nM was achieved. The proposed method has been successfully applied for clarithromycin content determination in spiked human plasma and urine samples [50]. Lincomycin was determined in the urine sample by microchip CE with integrated ITO working electrode based on ECL detection. This microchip CE–ECL system can be used for the rapid analysis of lincomycin within 40 s. This method is promising for detection of lincomycin in clinical and pharmaceutical area [51]. A novel method for ECL detection from successive electro- and chemo-oxidation of rifampicin was also reported and applied to the determination of rifampicin in pharmaceutical preparations and human urine. LOD (S/N = 3) was 3.9×10^{-8} mol L^{-1} [52].

Glyphosate (GLY) and (aminomethylphosphonic acid) AMPA were analyzed by CE–ECL detection method using ITO working electrode biased at 1.6 V [53]. Ionic liquid was used as the binder [54] and LODs for GLY and its major metabolite AMPA in water were 0.06 and 4.04 μg mL^{-1}, respectively. ECL detection along with CE separation was used for the rapid detection of spectinomycin for monitoring the drug in clinical and biochemical laboratories [55]. Using ultrasonic microdialysis, another more sensitive and selective CE–ECL method was used for determination of the number of binding sites and binding constant between diltiazem hydrochloride (DLT) and human serum albumin (HAS). Diltiazem hydrochloride, a calcium channel blocker, with linear range from 0.02 to 100 μmol L^{-1} ($r^2 = 0.9983$) and detection limit of 5.1 nmol L^{-1} was successfully developed [56].

Wang and his group also detected procyclidine (an anticholinergic drug) in human urine based on $Ru(bpy)_3^{2+}$ ECL method integrated with CE. ECL detection cell designed for this sensing method was based on post-column addition of $Ru(bpy)_3^{2+}$ and a detection limit of 1×10^{-9} mol L^{-1} in an on-capillary stacking mode was achieved. For detection of procyclidine in urine, a cartridge packed with slightly acidic cation-exchange resin was used to eliminate the matrix effects of urine and detection sensitivity was improved [57].

A simple and sensitive liquid chromatographic method coupled with ECL was developed for the separation and quantification of naproxen (a nonsteroidal anti-inflammatory drug) in human urine. The method was based on the ECL of naproxen in basic $NaNO_3$ solution with a dual-electrode system. The detection limit was 1.6×10^{-8} g mL^{-1} (S/N = 3) [58]. Furthermore, MEKC chromatography was used with ECL of $Ru(bpy)_3^{2+}$ as a fast and sensitive approach to detect an antipsychotic and antihypertensive drug, i.e., reserpine in urine. Field-amplified injection was used to minimize the effect of ionic strength in the sample and to achieve high sensitivity. In this way, the sample was analyzed directly without any pre-treatment with LOD (S/N = 3) to be 7.0×10^{-8} mol L^{-1} [59].

Pipemidic acid is extensively used in the treatment of gram-negative urinary tract infections, and the contents of proline in human urine vary in association with chronic uremia. Simultaneous determination of proline and pipemidic acid in human urine [60] and prolidase activity based on the determination of proline produced from enzymatic reaction [61] were performed using coupling of $Ru(bpy)_3^{2+}$-based ECL detection with CE. Numerous drugs are carboxylic acid derivatives containing amino group, and hydrolysis reaction of these agents often generates toxic amines. Thus, the detection of amine impurity is of great importance in drug quality control of these amino group-containing ester and amide. Another CE method coupled with end-column ECL based on $Ru(bpy)_3^{2+}$ system is used for the analysis of N,N-dimethyl ethanolamine, the degradation product of meclofenoxate, with a detection limit of 2.0×10^{-8} mol L^{-1} at the S/N = 3 [62]. For tricyclic antidepressants Imi and Tri, CE method was used in combination with

$Ru(bpy)_3^{2+}$ end-column ECL detection which was successfully applied for the determination of Imi in pharmaceutical dosage form. LODs of 5 nM and 1 nM (S/N = 3) were obtained for Imi and Tri, respectively [63].

A performant reagentless ECL system for H_2O_2 detection based on electropolymerized luminol on pre-treated screen-printed electrodes was developed [64]. An ECL biosensor based on carboxylic acid-functionalized multi-walled carbon nanotubes (COOH-F-MWNT) and Au nanoparticles [65] and immobilization in sol–gel hybrid material was fabricated for sensing the efficiency of ethanol. The intensity of ECL increased linearly with ethanol concentration from 2.5×10^{-5} to 5.0×10^{-2} M and detection limit was 1.0×10^{-5} M [66].

7.3   Life Sciences/Biomedical Analysis

Monosaccharides, an important class of carbohydrates, have been investigated using 2-diethylaminoethanethiol as a derivatizing reagent. Xylose, rhamnose, glucose, and glucosamine were selected as analytes. The method was successfully applied for the assay of glucose in angelica with a detection limit of 6.0×10^{-8} mol L^{-1} (S/N = 3) [67]. Immobilization of GOD onto a membrane-modified glassy carbon electrode, prepared by using poly(diallyldimethylammonium chloride) (PDDA) doped with chitosan, was used for ECL ultrasensitive and selective determination of glucose with good reproducibility and stability. Figure 7.2 shows scheme of the glucose ECL biosensor. Under the optimal conditions, the proposed ECL biosensor was able to detect glucose in the range of 0.5–4.0×10^4 nM with a detection limit of 0.1 nM [68].

Composite particles of Fe_3O_4/GOD were adhered by Qiang and his group onto solid paraffin carbon paste electrode surface by magnetic force to act as a working electrode. H_2O_2 was produced by enzymatic reaction of GOD, and ECL could be obtained by the reaction between luminol and H_2O_2 for sensing glucose [69].

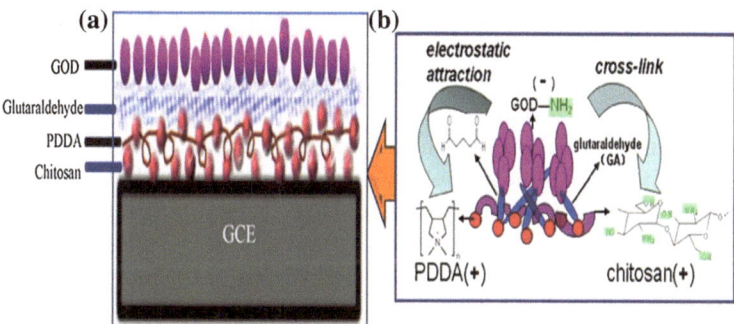

Fig. 7.2 Scheme of the glucose ECL biosensor. Adapted from Ref. [68]. Copyright 2009 Elsevier B.V

A FI optical fiber glucose and lactate biosensor [70–72] and a thin-film glucose ECL biosensor were also developed based on alcohol-free mesoporous molecular sieve silica modified electrode [73] and immobilization of GOD with MWNT/Nafion film [74–76].

A poly(nickel(II) tetrasulfophthalocyanine)/multi-walled carbon nanotubes composite modified electrode (polyNiTSPc/MWNTs) was fabricated by electropolymerization of NiTSPc on MWNTs-modified glassy carbon electrode on which GOD was immobilized to establish an ECL glucose sensor. This ECL sensor gave an outstanding well reproducibility and long-term stability with a detection limit of 8.0×10^{-8} mol L^{-1} (S/N = 3) [77]. Another sensor for the determination of H_2O_2, glucose, and GOD was established employing linear sweep voltammetry coupling with ECL technique. This sensor works with the ECL enhancement property of bis-[3,4,6-trichloro-2-(pentyloxycarbonyl)-phenyl]oxalate (BTPPO) at glassy carbon electrode in the presence of H_2O_2. The enhanced ECL intensity has a linear relationship with the concentration of glucose in the range of 1.0×10^{-4} to 1.0×10^{-3} mol L^{-1}, and the detection limit for glucose is found to be 5.0×10^{-5} mol L^{-1} (S/N = 3) [78].

Some studies reported the synthesis of poly(luminol–aniline) nanowires composite (PLANC) on the surface of graphite electrode by electro-oxidizing the mixture of luminol with aniline in the H_2SO_4 acidic medium to improve the analytical performances of the ECL biosensor for glucose [79]. Based on the linear relationship between concentration of H_2O_2 and the quenching of ECL in $Ru(bpy)_3^{2+}$/TPA system, procedures for the indirect determination of glucose with a FIA system were developed with a detection limit of 1.0 µM [80]. A multiparametric biochip for glucose, lactate, and choline determination has been developed based on luminol/H_2O_2 ECL. After optimizing the conditions, the ECL H_2O_2 sensor exhibited a detection limit of 2 mM and a working range from 2 to 0.5 mM. The multi-biosensor observed the detection of glucose, lactate, and choline in the ranges of 20–2 mM, 2–0.2 mM, and 2–0.2 mM, respectively [81]. A novel biosensor combining enzymatic selectivity with the sensitivity of ECL detection for quantification of enzyme substrate was described. Alcohol dehydrogenase (ADH) ECL biosensor by self-assembling ADH to $Ru(bpy)_3^{2+}$–AuNPs aggregates (Ru–AuNPs) on ITO electrode surface has been developed. This biosensor displayed wide linear range, high sensitivity, and good stability [82]. Ultrasensitive detection of antigen was obtained using ECL quenching by CdTe QDs through energy scavenging [83].

Based on its sensitizing effect on the weak ECL emission of electrochemically oxidized luminal, a sensitive and selective ECL-FI method for the determination of epinephrine, a neurotransmitter, was constructed. Under the optimum experimental conditions, the relative ECL intensity increased linearly with increasing epinephrine concentration in the range 7.0×10^{-8} to 6.0×10^{-6} mol L^{-1} and with a detection limit (S/N = 3) of 2.8×10^{-8} mol L^{-1} [84]. Another FI-ECL analysis method for the determination of L-lysine in alkaline Na_2CO_3-$NaHCO_3$ buffer solution, based on the enhanced ECL of $Ru(bpy)_3^{2+}$-L-lysine, was developed with a detection limit 1.43×10^{-9} g mL^{-1}. RSD for the determination of 1.0×10^{-6}

g mL^{-1} L-lysine was 1.8 % ($n = 11$) [85]. Further investigations revealed the studies on quantitative ECL detection of a model protein, BSA. The ECL was achieved via biotin–avidin interaction using an avidin-based, Ru(bpy)$_3^{2+}$-labeled ECL sensor. The proposed method was further applied successfully to the lysozyme with the detection limit of 0.1 mg L^{-1} [86]. Furthermore, BSA and IgG were determined by amplified ECL using 4-(Dimethylamino)butyric acid (DMBA) labeling combined with gold nanoparticles [87]. Quenching of Ru(bpy)$_3^{2+}$/TPA ECL system also leads to a highly sensitive approach for the detection of proteins, BSA and casein [88].

Ricin, a highly toxic protein presents in the seeds of Ricinus communis (castor), has been used as industrial lubricant, for poisoning, and to contaminate food. A monoclonal antibody-based method was developed for detecting and quantifying ricin in ground beef with a limit of detection 0.5 ng g^{-1} for the ECL method and 1.5 ng g^{-1} for enzyme-linked immunosorbent assay (ELISA) [89]. Choline has positive neuromuscular, metabolic, mental, and structural effects on the body of mammals, and it has recently been classified as a basic nutrient for humans by the Food and Nutrition Board of the Institute of Medicine (USA) [90]. It does not has ability to intensify ECL signals of Ru(bpy)$_3^{2+}$ to be detected. Rather when it is heated to its melting point, choline decomposes and forms two tertiary amines (trimethylamine and N, N-dimethyl-2-aminoethanol) that act as ECL co-reactants. This characteristic of choline is used to construct a simple biosensor for detection of choline in milk powder using ECL. The detection method overcame the defects of the commonly used method based on enzyme procedures and liquid chromatography–mass spectrometric technique [91].

Moreover, nanobiosensing of choline and acetylcholine was done by amplified ECL of QDs employing electrochemically reduced graphene oxide (ERGO). These biosensors were fabricated by covalently cross-linking ChO and AChE–ChO on QDs/ERGO/GCE which gave limit of detection 8.8 μM for choline and 4.7 μM for acetylcholine. This approach could be of potential applications in electronic device and bioanalysis [92]. The fact that choline cannot enhance the intensity of ECL signals of Ru(bpy)$_3^{2+}$ is used to construct a simple biosensor based on luminol ECL integrated in a FIA system; a fiber optic biosensor was developed for the detection of choline [93]. Other new biocompatible luminol ECL biosensor based on enzyme/titanate NTs/chitosan composite film (Fig. 7.3) [94, 95] and polymeric luminol as the luminophore instead of luminol in solution [96] with excellent stability and high sensitivity was developed for the determination of choline. In the fabrication of former ECL biosensor, biocompatible titanate nanotubes (TNTs) were immobilized on a chitosan-modified GCE via electrostatic adsorption. Other paper reported an FIA system to determine trace amount of choline in human urine. This method employs the coupling of an enzyme reactor with an ECL detector. This enzymatic FIA/ECL provided high sensitivity for the determination of choline with the detection limit as low as 0.05 μM [97].

For uric acid determination, an ECL disposable biosensor was constructed by immobilization in a double-layer design of luminol as a copolymer with 3,3′, 5,5′-tetramethylbenzidine (TMB) and the enzyme uricase in chitosan on gold

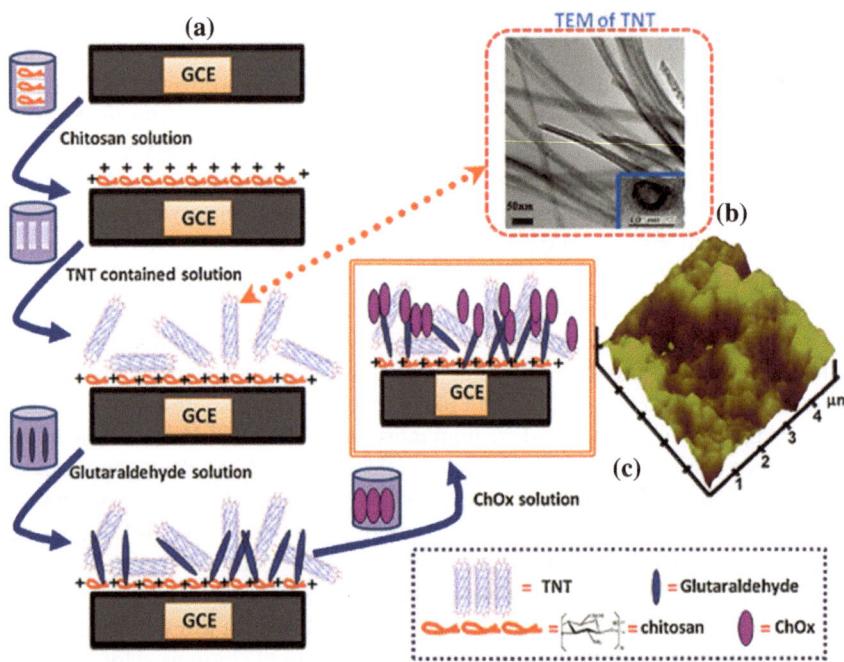

Fig. 7.3 A schematic steps for fabrication of the ECL biosensor (**a**). Insets are TEM (**b**) and AFM (**c**) images of TNTs and ChOx/TNTs/CHIT film. Adapted from Ref. [95]. Copyright 2009 Elsevier B.V

screen-printed cells. The improved electroluminescent characteristics of this copolymer poly(luminol–TMB) made it possible to determine uric acid by measuring the growing ECL emission with the analyte concentration. LOD of 4.4×10^{-7} M was obtained [98]. Another method for uric acid determination in human urine based on quenching of $Ru(bpy)_3^{2+}$ ECL was described by Zu [99]. A single-channel portable FI-ECL system for the determination of L-thyroxine based on its enhancement effect on $Ru(bpy)_3^{3+}$-NADH-ECL intensity was developed. The proposed method was applied to pharmaceutical preparations [100]. For the detection of tyrosine, a sulfite-enhanced anodic ECL method of MPA-modified CdTe QDs in air-saturated aqueous solution based on quenching of ECL emission has been described [101].

ECL-FI method was used for noradrenaline and DA determination [102]. This method was practically used to determine the neurotransmitters in cerebrospinal fluid (CSF) with DA as the index [103]. Two maximal potential-resolved FI-ECL peaks were observed for $Ru(bpy)_3^{2+}$/TPA system at 0.90 and 1.05 V, and for $Ru(phen)_3^{2+}$/TPA at 1.01 and 1.25 V (vs. Ag/AgCl) in pH 8.0 phosphate buffer solutions. Sensitive ECL inhibition effects were observed in the presence of noradrenaline and DA for both of these systems. Another simple and sensitive FI method has been developed for the determination of DA based on its inhibition of the ECL of luminal with a detection limit of 30 nmol L^{-1} [104]. Lin found DA to be able to inhibit the ECL of $Ru(bpy)_3^{2+}$/DBAE system. Thus, the new $Ru(bpy)_3^{2+}$/DBAE-DA ECL system seemed to have great potential for its application in analytical chemistry. Under the optimum conditions, the logarithmic plot of the inhibited ECL versus the concentration of DA was linear in the range of 5×10^{-10} to 7×10^{-7} M and the detection limit of 4.0×10^{-11} M was observed [105]. Furthermore, sensitive ECL sensor based on the modification of composite of QDs, CNTs, and chitosan on ITO glass was fabricated. The developed sensor displays efficient and stable anodic ECL which is quenched by DA [106].

Based on ECL inhibition, DA and epinephrine were separated by CE and indirectly determined. Linear ranges for both analytes were 0.1–10 mM and the detection limits (S/N = 3) were 10 nM for DA and 30 nM for epinephrine [107]. A selective and sensitive sensor based on dual-cloud point extraction (DCPE) and tertiary amine labeling for the quantification of indole-3-acetic acid (IAA) and indole-3-butyric acid (IBA) by CE-ECL has been demonstrated. The detection limits (S/N = 3) were 2.5 and 2.8 nM for IAA and IBA, respectively. The proposed method was applied successfully to the detection of low concentrations of two auxins in acacia tender leaves, buds, and bean sprout [108].

(Dimethylamino)ethanethiol DMAET-capped CdTe NPs were employed in the construction of aptamer-based biosensing system for proteins [109]. Highly efficient and sensitive determination of alkaloids and amino acids using CE in conjunction with sequential light-emitting diode-induced fluorescence (LEDIF) and ECL method was carried out. In the CE-LEDIF-ECL system, the ECL detector was located in the outlet of the capillary, while the LEDIF detector was positioned 12 cm from the outlet [110]. $Ru(bpy)_3^{2+}$ in MWCNTs/Nafion composite membrane was immobilized for an ECL method for reduced nicotinamide adenine dinucleotide (NADH) with a detection limit of 8.2×10^{-7} M [111].

7.4 Environmental Analysis

ECL in combination with some techniques has also been used to determine pesticides [112], herbicides [113], and alkaloids. Pesticides [114], acephate and dimethoate, were determined using a composite electrode (a TBR–multi-walled carbon nanotube paste electrode) with coreagent TPA [115]. Other literature revealed a rapid method for the determination of galanthamine (in Bulbus lycoridis radiatae) [116]. For sensitive determination of verticine and verticinone in Bulbus fritillariae, CE/$Ru(bpy)_3^{2+}$ ECL system with the assistance of ionic liquids was successfully established. Detection limits of 1.25×10^{-10} mol L^{-1} for verticine and 1×10^{-10} mol L^{-1} for verticinone were obtained (S/N = 3). Developed method was successfully applied to determine the amounts of alkaloids in Bulbus fritillariae [117].

For a nephrotoxic toxin, ochratoxin A (OTA), an ECL biosensor based on DNA aptamer as the recognition element and N-(4-aminobutyl)-N-ethylisoluminol

(ABEI) as the signal-producing compound was constructed. The relative standard deviation was 3.8 % at 0.2 ng mL^{-1} ($n = 7$). A detection limit of 0.007 ng mL^{-1} was obtained with the designed sensor. The proposed analytical method has been applied to measure OTA in naturally contaminated wheat samples [118]. ECL detection with CE separation system was used for the rapid analysis of nicotine in urine and cigarette samples [119], and pseudolycorine in the bulb of lycoris radiata (based on ultrasonic-assisted extraction) [120].

Wang [121] and You [122] groups demonstrated a rapid and simple analytical method for determination of alkaloids (atropine, anisodamine, and scopolamine) in Flos daturae extract by NACE coupled with ECL and EC dual detection was developed. A short capillary of 18 cm was used, and the decoupler was not needed. Electrophoretic buffer used was the mixture of acetonitrile (ACN) and 2-propanol containing Macetic acid (HAc), sodium acetate (NaAc), and tetrabutylammonium perchlorate (TBAP). Trimethylamine in fish [123] was quantitatively determined using a platinum electrode at the potential 1.23 V versus an Ag–AgCl electrode in a 10 mM solution of sodium borate (pH 9.2) containing 33 mM of TBR. A detection limit of 10^{-8} M was observed. Some other data were published on the highly sensitive determination of malachite green residues in fish, which was performed by using an ECL inhibition method combined with MISPE and a detection limit of about 6 ppt was obtained. This method was based on quenching the ECL signal of luminol by MG [124].

7.5 DNA Detection and Quantification

Most of the recent DNA hybridization detection methods employ the use of assay format and sandwich format assays as well. (Fig. 7.4a, b) [125–127]. In the largest part of publications, Ru(bpy)$_3$$^{2+}$ or its derivatives are widely used as ECL labels and TPA has been employed as a co-reactant. While some papers have also used several dsDNA intercalators and oxalate as reductants of electrogenerated Ru(bpy)$_3$$^{3+}$ [128] and co-reactant for DNA detection [129]. Different strategies have been employed for the sake of improving sensitivity; such as loading or immobilization of multiple ECL labels in microsized polystyrene microspheres [130] and CNTs [127] or Au NPs [131]. As for immobilization substrates of DNA, besides millimeter-sized Au electrodes, micrometer-sized Au chips [132] and anodically oxidized GC electrodes [133] have also been employed.

Combination of assembled microcell [134] or Au chip electrode [132] with a positive-intrinsic-negative (PIN) photodiode detector lead to advancement in the miniaturization of ECL instrumentation for DNA quantification. Though, noise level and sensitivity of silicon PIN diode are less sensitive than a PMT [134]. A highly sensitive DNA biosensor, fabricated by self-assembling the ECL probe of thiolated hairpin DNA tagged with ruthenium complex on the surface of a gold electrode works on a "switch-off" mode and ECL intensity decreases with an increase of the concentration of target DNA, and a detection limit of 9×10^{-11} M

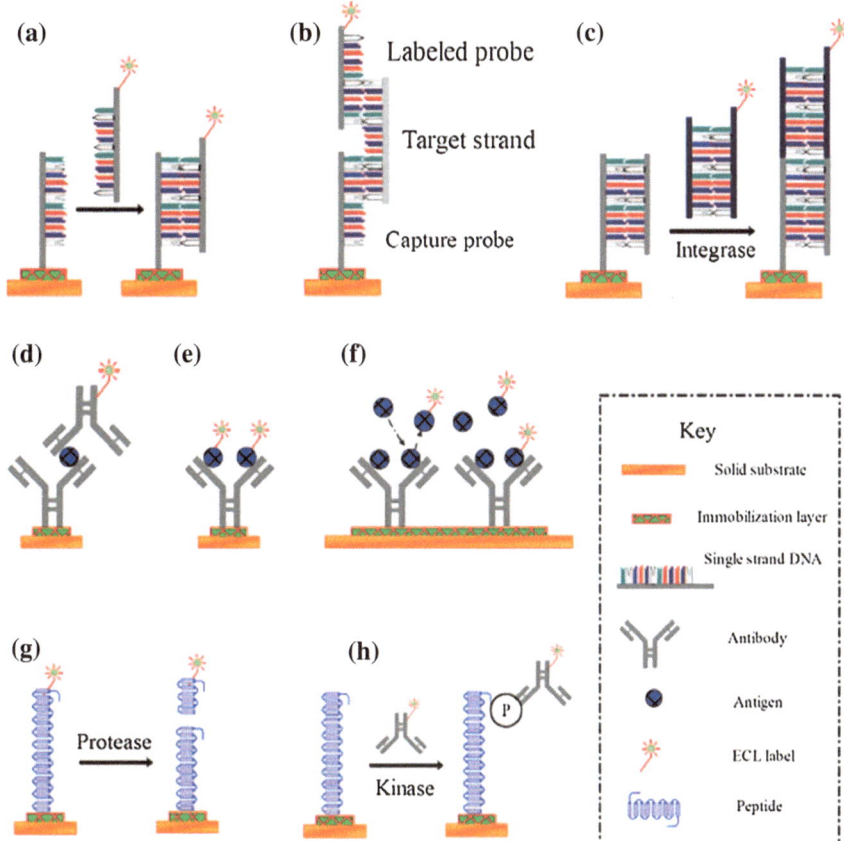

Fig. 7.4 Examples of solid-phase ECL assay formats. **a** DNA hybridization assay based on an immobilized ssDNA hybridizes with a labeled target ssDNA. **b** Sandwich-type DNA biosensor. **c** Assay used for integrase activity test with immobilized and free-labeled dsDNA. **d** Sandwich-type immunoassay. **e** Direct immunoassay. **f** Competitive assay in which analyte competes with labeled analyte for antibody-binding sites on immobilized antibody. **g** Protease activity assay in which cleavage of the immobilized peptide results in the decrease in ECL emission due to the removal of the ECL label. **h** Kinase activity assay using a labeled antibody to recognize the phosphorylated product. Adapted from Ref. [23]. Copyright 2008 American Chemical Society

complementary target ss-DNA was achieved [135]. ssDNA-Ru(bpy)$_3$$^{2+}$ upon its hybridization with the second ssDNA tagged with Cy5 dye results in an efficient ECL quenching and is found to be another promising approach for DNA detection and quantification [136]. Besides this, other researchers found application of quenching effect of Fc for DNA hybridization. This biosensor is based on intrinsic cathodic ECL at thin-oxide-covered glassy carbon (C/C$_x$O$_{1-x}$) electrodes [137] and Fc-MB [138]. Great advantage of the biosensor is attributed to its simplicity and cost-effectiveness with ECL generated from the electrode itself.

In another study, DNA-mismatched discrimination was revealed to be considerably better when a negative electrode potential of -0.3 V vs Ag was applied for 300 s prior to the ECL detection [139]. Table 7.1 lists some recent examples of DNA hybridization detection and quantification on the basis of $Ru(bpy)_3^{2+}$ ECL [23].

A sandwich assay format based on polymerase chain reactions (PCR)-amplified DNA strands immobilization on the surface of streptavidin-coated MB via biotin-labeled reverse primers, and the other ends linked with $Ru(bpy)_3^{2+}$ molecules was constructed for DNA fragments detection. In the proposed sensor, the PCR-amplified DNA strands were typically immobilized on the surface of streptavidin-coated MB via biotin-labeled reverse primers and the other ends were linked with $Ru(bpy)_3^{2+}$ molecules. Highly sensitive methods are needed for the detection of PCR products to reduce the number of amplification cycles because before the sufficient generation of product; errors are exponentially amplified through multiple cycles toward a plateau phase [23]. ECL detection method combined with bead-based PCR-amplified DNA-aptamers has been utilized for the detection of anthrax spores [141]. Bead-based PCR-ECL system has also been applied to the detection of various genes from viruses and diseases in human blood and serum [142–144] and plants [145], genetically modified organisms such as vegetables [146], and the measurement of DNA helicase activity of *Escherichia coli* DNA [147].

Luminol-H_2O_2-based method [148, 149] and hot electron-induced ECL [150] have also been reported. In the former case, immobilization of N-(4-aminobutyl)-N-etylisoluminol (ABEI), a luminol derivative, on a target ssDNA was employed for the ECL detection of the hybridization between a complementary probe ssDNA covalently linked to a polypyrrole support and the target ssDNA [149]. In hot electron-induced ECL system, hybridization was detected using hot electron-induced ECL of the tetramethylrhodamine dye. The proposed system employs thin-oxide film-coated Al and Si electrodes modified with an aminosilane layer and derivatized with short, 15-mer DNAs via diisothiocyanate coupling. Target DNAs were conjugated with tetramethylrhodamine dye at their amino-modified $5'$ end, detecting hybridization using hot electron-induced ECL of the dye with LOD in picomolar levels. Recently, catalytic guanine and adenine bases oxidation using $Ru(bpy)_3^{2+}$-modified GC electrode with a cast carbon nanotube/Nafion/$Ru(bpy)_3^{2+}$ composite film has also been employed for label-free ECL DNA detection [151]. ECL signals of dsDNA and its thermally denatured counterparts were distinctly discriminated at a low concentration of 30.4 nM for salmon testes DNA, and single-base mismatch detection of p53 gene sequence segment was realized with 0.4 nM.

Previous studies reveal the direct ECL detection of DNA in poly(vinylpyridine) (PVP) ultrathin film using cationic polymer $[Ru(bpy)_2(PVP)_{10}]^{2+}$ or $[Os(bpy)_2(PVP)_{10}]^{2+}$ [152]. DNA damage from benzo[a]pyrene metabolites using ECL arrays suitable for genotoxicity screening was also demonstrated later [23].

Table 7.1 Applications of $Ru(bpy)_3^{2+}$ for the detection of DNA hybridization

DNA sequence[a]	Detection method	Comments[d]	Refs.
5'-SHC6-22-mer[b]	Using DNA-binding intercalator as a reductant of electrogenerated $Ru(bpy)_3^{3+}$ with a throughhole type DNA biosensor	dsDNA intercalated by doxorubicin, daunorubicin, and 4,6-diamidino-2-phenylindole (DAPI), in which DAPI showed efficient ECL	[128]
5'-NH2-24-mer	$Ru(bpy)_3^{2+}$-doped silica NPs as ECL labels using assay format, ECL co-reactant: 2.5 mM H2C2O4 in PBS, pH 6.6	LR = 0.20 pM–2.0 nM, LOD = 0.10 pM, three-base mismatch and noncomplementary sequences showed almost no ECL	[129]
5'-21-mer	C6NH2SH, 5'-NH2C6-18-mer, and 42-mer	carbon nanotubes-loaded $Ru(bpy)_3^{2+}$ as ECL labels using sandwich-type DNA sensor, ECL in 0.10 M PBS (pH 7.4)–0.10 M TPrA	LR)
24 fM–1.7 pM, LOD) 9.0 fM,	discriminated two-base mismatched ssDNA	[127]	
5'-NH2C6 23-mer[c]	Au(111)/SAM-DNA using assay format with $Ru(bpy)_3^{2+}$ tag in 0.10 M LiClO4-0.10 M Tris-0.10 M TPrA (pH 8.0)	A series of electrode treatments (e.g., blocking free-COOH groups and pinholes, washing, and inert gas spraying) were used to reduce the nonspecific adsorption of the labeled species	[140]
5'-biotin-TEG 23-mer	Using $Ru(bpy)_3^{2+}$-loaded PSB as ECL labels, hybridized DNA separated magnetically, ECL in MeCN-0.055 M TFAA-0.10 M TPrA-0.10 M (TBA)BF4	LR = 1.0 fM–10 nM, distinguished two-base-pair mismatched from noncomplementary DNA hybridization	[130]
5' 18-mer-C3 SH-3'	Target ssDNA immobilized on Au electrode hybridized with probe ssDNA labeled with $\sim$15 nm Au NPs covalently attached with a large number of $Ru(bpy)_3^{2+}$, ECL in 0.10 M PBS (pH 7.4)-0.10 M TPrA	LR = 10 pM–10 nM, LOD = 5.0 pM	[131]
5'-SH (or NH2)-19-mer	Hybridization on integrated Au chip, light detection with a PIN photodiode detector, assay was based on displacement of DNA hybridization, ECL in phosphate buffer (pH 8.0)-0.10 TPrA	$Ru(bpy)_3^{2+}$ LOD = 0.1 fM	[132]
15-mer	3' end of one ssDNA labeled with $Ru(bpy)_3^{2+}$ hybridized with another ssDNA tagged with Cy5 at 5' end, efficient ECL quenching was observed in $\sim$0.3 M PBS (pH 7.5)–0.1 M TPrA-SDS (<0.1 % w %)	LOD = 30 nM, quenching efficiency = 78 %	[136]

(continued)

Table 7.1 (continued)

DNA sequence[a]	Detection method	Comments[d]	Refs.
5'-NH2-15-mer	Ru(bpy)$_2$(phen)$^{2+}$ derivative as ECL label, assay format was used, considerably better discrimination of two-base-pair mismatch was obtained with −0.3 V vs Ag for 300 s, Au chip electrode was used, ECL in 0.3 M PBS (pH 7.8)–0.1 M TPrA-0.1 % SDS	LOD = 1 pM in 30 μL of buffer	[139]
5'-NH2-18-mer	Anodically oxidized GC electrode was used for covalent immobilization of DNA, hybridization detection using labeled Ru(bpy)$_3$$^{2+}$ was conducted in PBS buffer-4 % TPrA	LOD < 1 pM	[133]

[a] Synthetic DNA sequences were used, unless otherwise stated. [b] Target organism, hepatitis A, B, and C virus, [c] Target organism, *Bacillus anthracis* (Ba813). [d] *LR*, linear range; *LOD*, limit of detection

7.6 Aptamer-Based Sensors

Aptamers are oligonucleic acid or peptide molecules that can selectively bind a specific target molecule. Recently, for the determination of a small-molecule drug, a highly sensitive ECL aptamer-based biosensor has been constructed based on two components: a cocaine-binding aptamer as molecular recognition element and as a model analyte and a Ru(II) complex as an ECL label [153].

A gold electrode immobilized with the ECL probe, which was a 5'-terminal cocaine-binding aptamer with the ECL label at 3'-terminus of the aptamer, upon recognition of the target cocaine generated an enhanced ECL signal. A change in the conformation of the ECL probe from random coil-like configuration on the probe-modified film to three-way junction structure, in close proximity to the sensor interface lead to the enhancement of ECL signal with a detection limit of 1.0 nM. A novel ECL biosensor based on the construction of triplex DNA for the detection of adenosine (a purine nucleoside) was designed, comprised of an aptamer as a molecular recognition element, and quenches ECL of Ru(bpy)$_3$$^{2+}$ by Fc mono-carboxylic acid. The presence of adenosine, the aptamer sequence (Ru-DNA-1), prefers to form the aptamer–adenosine complex with hairpin configuration, and the switch of the DNA-1 occurs in conjunction with the generation of a strong ECL signal owing to the dissociation of a quenching probe. The detection limits were 2.7×10^{-10} mol L^{-1} and 2.3×10^{-9} mol L^{-1} for the ECL-triplex DNA biosensor and ECL-duplex DNA biosensor, respectively [154].

Another construction procedure of Fc-labeled structure-switching signaling aptamer as a solid-state ECL sensing platform for detection of adenosine provided a promising method for aptamer-based small-molecule detection due to its simplicity, rapidity, low cost, and excellent response characteristics [155].

It has already been reported that Fc has the ability to efficiently and stably quench the ECL of $Ru(bpy)_3^{2+}$ at the electrode. Based on this characteristic of Fc, a novel ECL aptamer biosensor for detection of adenosine has been fabricated. $Ru(bpy)_3^{2+}$-doped silica nanoparticle (Ru-SiNP) works as a novel biocompatible ECL tag material, and its ECL is quenched by Fc. The biosensor offers several advantages, such as good selectivity, high sensitivity, reproducibility, and stability. Additionally, the sensor could be easily regenerated and reused at least 30 times, which would save time and chemical reagent [156]. Using cDNA oligonucleotide as sensing probe and aptamer as recognition element trace amount of ATP is detected. The ECL-AB biosensor combined both highly selective ATP-binding aptamer and highly sensitive ECL technique. This biosensor design has several unique features such as cost-effectiveness, practical reusability, high sensitivity, and high selectivity [157].

A highly sensitive detection method for thrombin on the basis of ECL quenching through capture of ferrocene-labeled ligand-bound aptamer molecular beacon (MB) [158] and efficient energy-transfer-induced ECL quenching from CdS:Mn nanocrystals film to CdTe QDs-doped silica nanoparticles (CdTe/SiO2 NPs) was established [159]. This aptasensor showed a detection limit of 1.7 pM and 1 aM, respectively. Based on structure-switching ECL-quenching mechanism, a novel signal-on junction-probe ECL aptamer biosensor using complex of Au nanoparticle and ruthenium (II) tris-bipyridine ($Ru(bpy)_3^{2+}$–AuNPs) on the surface of gold electrode has been developed for the detection of ultratrace thrombin with a detection limit of 8.0×10^{-15} M [160]. Another application of quenching effect of ferrocene was developed by fabricating the solid-state ECL protein biosensor based on the competing and substitute reaction between protein-to-DNA aptamer and DNA-to-DNA aptamer [161].

Recently, another application of quenching effect of Fc was developed by fabricating a solid-state ECL biosensing switch for the detection of T4 DNA ligase [162]. This bioassay system was based on special Fc-labeled molecular beacon (Fc-MB) and consisted of an ECL substrate and an ECL intensity switch. Some structural changes happen during the reaction, resulting in an obvious increment in ECL intensity due to the decreased Fc quenching effect to the ECL substrate.

Besides for the detection of proteins and small biomolecules, the combination of ECL and aptamers has also been used for a high-throughput assay of an RNA editing reaction [163]. RNA editing is the molecular process in which the information content in an RNA molecule is altered through a chemical change in the base makeup [164]. The ECL assay is performed within streptavidin-coated microtiter plates that have carbon electrodes running through the bottom of the plates. The reporter RNA is labeled with a Ru complex that can generate ECL only when placed close to these electrodes. Editing leads a conformational change in the reporter that activates a streptavidin-binding aptamer. This results in the immobilization of the RNA at the bottom of the microliter plates and the generation of ECL. The assay is sensitive to low-femtomole quantities of edited products and suites for high-throughput drug screening.

7.7 Food and Water Safety and Military/Defense Applications

Foodborne pathogens are one of the causes of food contamination leading to the disease outbreaks in public health. Hence, tests for these pathogens are important tools in the food industry in screening products. Though current culture methods take long time (2–4 days) to detect pathogens in the low concentrations required for infectivity; ECL-based assays provide a fast and alternative testing method and are ideal for food testing because of their high tolerance to various matrices.

Magnetic bead separation played an important role in improving the detection sensitivity; ECL coupled with magnetic bead separation was used to develop assays for a variety of biotoxoids that are important in many fields, for environment, food, and water industries. For few years, magnetic bead separation coupled with ECL technique is also gaining considerable importance in military applications. ECL is a very sensitive technique; thus, ECL combined with other techniques improves the sensitivity and detection limit. For example, standard radioimmunoassays do not show detection limit as low as the detection limit achieved by assays for bovine luteinizing hormone with ECL [165].

Several papers also reported highly sensitive bacterial and viral assays involving ECL [166–168] as well as cancer antigens [169–172] in a variety of templates employing immunoassays and nucleic acid amplification techniques. *E. coli* 0157 studies have been done for water safety and food, such as drinking [173] and creek [174] water, feces [175], various food and environmental water matrices [176–178], and ground beef, chicken, fish, juices, and milk [179, 180]. In recent years, an assay for *E. coli* that lacks the need of culturing step was demonstrated for water samples based on a mRNA sequence-coding detection for a specific heat shock protein [181]. An assay sensitivity equal to or greater than conventional assays using flow cytometry, ELISA, and radioallergosorbent test (RAST) was obtained for some other species such as Cryptosporidium parvum oocysts [173, 174, 182, 183], *Campylobacter, Listeria monocytogenes, Salmonella*, and *Staphylococcus aureus* enterotoxins [176, 179, 184–186].

Several biological threat agents were also detected employing ECL approach including purified Staphylococcus aureus enterotoxin B, botulinus A, cholera toxin, and ricin toxins [187]. Another study in the following years focused on *Bacillus anthracis* and *E. coli* O157:H7 and used a male-specific coliphage virus to stimulate the detection of viral agents such as smallpox, Ebola, and yellow fever [186]. This work was later on extended to "field" samples to measure

Staphylococcus aureus enterotoxin B in serum, tissue, and urine, while anthrax to measure in saliva swabs [178] and soil samples [177].

A significant application of ECL in military is the detection of explosives or explosives' degradation products. For example, diaminotoluene isomers (often associated with the degradation of explosives such as TNT) in the presence of Au^+ and Cu^{2+} ions form weakly ECL compounds [141]. This approach may find use in military applications. Interestingly, an aptamer-magnetic particle ECL has been used in combination with systematic evolution of ligands by exponential enrichment (SELEX). Highly specific receptor ligands (known as aptamers) were used with varying affinities to anthrax spores. Aptamers have potential for use as inexpensive, in vitro-generated receptors for biosensors in biological warfare detection and other areas [188].

With the increased emphasis on detecting biological threat agents, this area will undoubtedly expand in coming years [189]. Increasing contribution of Igen International with various US government agencies and programs furnishes another proof of the efficacy of ECL in real diagnostic field, aiming at mounting detection systems for homeland security and military defense [190]. Numerous series within the department of defense for instruments (designed for battlefield use or deployment within mobile laboratories throughout the combat zone), reagents, and assay development for the detection of biological agents or toxins in environmental and clinical samples are running. Another agreement of cooperative research and development (dealing with the developing tests for food-, water-, and environmentally borne toxins) with the US Army Medical Research Institute of Infectious Diseases (USAMRIID) is also operating successfully to shield American troops from diseases caused by biological agents, and the tests are designed for use by government agencies, food processors, and analytical laboratories [176].

References

1. Ludvik J (2011) DC-electrochemiluminescence (ECL with a coreactant)-principle and applications in organic chemistry. J Solid State Electrochemy 15(10):2065–2081. doi:10.1007/s10008-011-1546-x
2. Hu L, Xu G (2010) Applications and trends in electrochemiluminescence. Chem Soc Rev 39(8):3275–3304. doi:10.1039/b923679c
3. Bard AJ, Debad JD, Leland JK, Sigal GB, Wilbur JL, Wohlsatdter JN (2000) Encyclopedia of Analytical Chemistry: Applications, Theory and Instrumentation, vol 11. John Wiley & Sons Inc, New York
4. Cruser SA, Bard AJ (1967) Concentration-intensity relationships in electrogenerated chemiluminescence. Anal Lett 1(1):11–17
5. McCord P, Bard AJ (1991) Electrogenerated chemiluminescence: Part 54. Electrogenerated chemiluminescence of ruthenium(II) 4,4'-diphenyl-2,2'-bipyridine and ruthenium(II) 4,7-diphenyl-1,10-phenanthroline systems in aqueous and acetonitrile solutions. J Electroanal Chem Interfacial Electrochem 318(12):91–99

6. Miao W, Bard AJ (2004) Electrogenerated chemiluminescence. 80. C-reactive protein determination at high amplification with [Ru(bpy)$_3$]$^{2+}$-containing microspheres. Anal Chem 76(23):7109–7113. doi:10.1021/ac048782s

7. Lu Y, Young J, Meng YG (2007) Electrochemiluminescence to detect surface proteins on live cells. Curr Opin Pharmacol 7(5):541–546

8. Kurita R, Arai K, Nakamoto K, Kato D, Niwa O (2010) Development of electrogenerated chemiluminescence-based enzyme linked immunosorbent assay for sub-pM detection. Anal Chem 82(5):1692–1697

9. Wu Y, Shi H, Yuan L, Liu S (2010) A novel electrochemiluminescence immunosensor via polymerization-assisted amplification. Chem Comm 46(41):7763–7765. doi:10.1039/c0cc02741c

10. Gan N, Hou J, Hu F, Cao Y, Li T, Guo Z, Wang J (2011) A renewable and ultrasensitive electrochemiluminescence immunosenor based on magnetic RuL@SiO$_2$-Au similar to RuL-Ab$_2$ sandwich-type nano-immunocomplexes. Sensors 11(8):7749–7762. doi:10.3390/s110807749

11. Qian J, Zhou Z, Cao X, Liu S (2010) Electrochemiluminescence immunosensor for ultrasensitive detection of biomarker using Ru(bpy)$_3$$^{2+}$-encapsulated silica nanosphere labels. Anal Chim Acta 665(1):32–38. doi:10.1016/j.aca.2010.03.013

12. Wilson R, Barker MH, Schiffrin DJ, Abuknesha R (1997) Electrochemiluminescence flow injection immunoassay for atrazine. Biosens Bioelectron 12(4):277–286. doi:10.1016/s0956-5663(96)00067-x

13. Wu A-H, Sun J-J, Fang Y-M, Su X-L, Chen G-N (2010) Hot electron induced cathodic electrochemiluminescence at AuSb alloy electrode for fabricating immunosensor with self-assembled monolayers. Talanta 82(4):1455–1461. doi:10.1016/j.talanta.2010.07.017

14. Arai K, Takahashi K, Kusu F (1999) An electrochemiluminescence flow through-cell and its applications to sensitive immunoassay using N-(aminobutyl)-N-ethylisoluminol. Anal Chem 71(11):2237–2240. doi:10.1021/ac9810361

15. Jie G, Liu P, Wang L, Zhang S (2010) Electrochemiluminescence immunosensor based on nanocomposite film of CdS quantum dots-carbon nanotubes combined with gold nanoparticles-chitosan. Electrochem Comm 12(1):22–26. doi:10.1016/j.elecom.2009.10.027

16. Marquette CA, Blum LJ (1998) Electrochemiluminescence of luminol for 2,4-D optical immunosensing in a flow injection analysis system. Sens Actuators, B 51(1–3):100–106. doi:10.1016/s0925-4005(98)00175-0

17. Sun S, Yang M, Kostov Y, Rasooly A (2010) ELISA-LOC: lab-on-a-chip for enzyme-linked immunodetection. Lab Chip 10(16):2093–2100. doi:10.1039/c003994b

18. Gan N, Hou J, Hu F, Cao Y, Li T, Zheng L, Wang J (2011) Sandwich-type electrochemiluminescence immunosensor based on PDDA-G@Lu-Au composite for alpha-fetoprotein detection. Int J Electrochem Sci 6(11):5146–5160

19. Wilson R, Clavering C, Hutchinson A (2003) Electrochemiluminescence enzyme immunoassay for TNT. Analyst 128(5):480–485. doi:10.1039/b301942j

20. Wilson R, Clavering C, Hutchinson A (2003) Paramagnetic bead based enzyme electrochemiluminescence immunoassay for TNT. J Electroanal Chem 557:109–118. doi:10.1016/s0022-0728(03)00353-x

21. Egashira N, S-i Morita, Hifumi E, Mitoma Y, Uda T (2008) Attomole detection of hemagglutinin molecule of influenza virus by combining an electrochemiluminescence sensor with an immunoliposome that encapsulates a Ru complex. Anal Chem 80(11):4020–4025

22. Wang X, Bobbitt DR (1999) In situ cell for electrochemically generated Ru(bpy)$_3$$^{3+}$-based chemiluminescence detection in capillary electrophoresis. Anal Chim Acta 383(3):213–220

23. Miao W (2008) Electrogenerated chemiluminescence and its biorelated applications. Chem Rev 108(7):2506–2553. doi:10.1021/cr068083a

24. Lin JM, Yamada M (1998) Electrogenerated chemiluminescence of methyl-9-(p-formylphenyl) acridinium carboxylate fluorosulfonate and its applications to immunoassay. Microchem J 58(1):105–116. doi:10.1006/mchj.1997.1539

25. Xu XHN, Jeffers RB, Gao JS, Logan B (2001) Novel solution-phase immunoassays for molecular analysis of tumor markers. Analyst 126(8):1285–1292. doi:10.1039/b104180k

26. Qian J, Zhang C, Cao X, Liu S (2010) Versatile immunosensor using a quantum dot coated cilica nanosphere as a label for signal amplification. Anal Chem 82(15):6422–6429. doi:10.1021/ac100558t

27. Wei H, Liu J, Zhou L, Li J, Jiang X, Kang J, Yang X, Dong S, Wang E (2008) Ru(bpy)$_3^{2+}$-doped silica nanoparticles within layer-by-layer biomolecular coatings and their application as a biocompatible electrochemiluminescent tag material. Chem Eur J 14(12):3687–3693. doi:10.1002/chem.200701518

28. Wu YY, Li T, Liang H, Xue J (2005) Separation and determination of bupivacaine in plasma by capillary electrophoresis with tris(2,2′-bipyridyl)ruthenium(II) electrochemiluminescence detection. Luminescence 20(4–5):352–357. doi:10.1002/bio.855

29. Xu YH, Gao Y, Wei H, Du Y, Wang EK (2006) Field-amplified sample stacking capillary electrophoresis with electrochemiluminescence applied to the determination of illicit drugs on banknotes. J Chromatogr A 1115(1–2):260–266. doi:10.1016/j.chroma.2006.02.084

30. Han B, Du Y, Wang E (2008) Simultaneous determination of pethidine and methadone by capillary electrophoresis with electrochemiluminescence detection of tris(2,2′-bipyridyl)ruthenium(II). Microchem J 89(2):137–141. doi:10.1016/j.microc.2008.01.007

31. Pan W, Liu YJ, Huang Y, Yao SZ (2006) Determination of difenidol hydrochloride by capillary electrophoresis with electrochemiluminescence detection. J Chromatogr B 831(1–2):17–23. doi:10.1016/j.jchromb.2005.11.020

32. Liu JF, Cao WD, Qiu HB, Sun XH, Yang XR, Wang EK (2002) Determination of sulpiride by capillary electrophoresis with end-column electrogenerated chemiluminescence detection. Clin Chem 48(7):1049–1058

33. Li JG, Zhao FJ, Ju HX (2006) Simultaneous electrochemiluminescence determination of sulpiride and tiapride by capillary electrophoresis with cyclodextrin additives. J Chromatogr B 835(1–2):84–89. doi:10.1016/j.jchromb.2006.03.017

34. Liu Y-M, Shi Y-M, Liu Z-L, Tian W (2010) A sensitive method for simultaneous determination of four macrolides by CE with electrochemiluminescence detection and its applications in human urine and tablets. Electrophoresis 31(2):364–370. doi:10.1002/elps.200900302

35. Yang R, Zeng H-J, Li J–J, Zhang Y, Li S-J, Qu L-B (2011) Capillary electrophoresis coupled with end-column electrochemiluminescence for the determination of ephedrine in human urine, and a study of its interactions with three proteins. Luminescence 26(5):374–379. doi:10.1002/bio.1336

36. Liu YM, Peng LF, Mei L, Liu LJ (2011) Determination of diastereoisomeric alkaloids in urine by capillary electrophoresis with electrochemiluminescence detection. Chin Chem Lett 22(2):197–200. doi:10.1016/j.cclet.2010.10.019

37. Huang Y-S, Chen S-N, Whang C-W (2011) Capillary electrophoresis-electrochemiluminescence detection method for the analysis of ibandronate in drug formulations and human urine. Electrophoresis 32(16):2155–2160. doi:10.1002/elps.201100202

38. Zheng XW, Zhang ZJ, Li BX (2001) Flow injection chemiluminescence determination of captopril with in situ electrogenerated Mn^{3+} as the oxidant. Electroanalysis 13(12):1046–1050. doi:10.1002/1521-4109(200108)13

39. Zheng XW, Qu YJ, Zhang ZJ, Zhang CM (2005) Highly sensitive electrogenerated chemiluminescence detecting ranitidine based on chemically modifying microenvironment of the chemiluminescence reaction. Electroanalysis 17(11):1008–1014. doi:10.1002/elan.200403210

40. Michel PE, Fiaccabrino GC, de Rooij NF, Koudelka-Hep M (1999) Integrated sensor for continuous flow electrochemiluminescent measurements of codeine with different ruthenium complexes. Anal Chim Acta 392(2–3):95–103

41. Liu YJ, Pan W, Liu Q, Yao SZ (2005) Study on the enhancement of Ru(bpy)$_3^{2+}$ electrochemiluminescence by nanogold and its application for pentoxyverine detection. Electrophoresis 26(23):4468–4477. doi:10.1002/elps.200500391

42. Deng B, Yin H, Liu Y, Ning X (2011) Pharmacokinetics of propranolol hydrochlorid in human urine by capillary electrophoresis coupled with electrochemiluminescence. Anal Sci 27(1):55–59. doi:10.2116/analsci.27.55

43. Huang J, Sun J, Zhou X, You T (2007) Determination of atenolol and metoprolol by capillary electrophoresis with Tris(2,2'-bipyridyl)ruthenium(II) electrochemiluminescence detection. Anal Sci 23(2):183–188. doi:10.2116/analsci.23.183

44. Wang Y, Wu Q, Cheng M, Cai C (2011) Determination of beta-blockers in pharmaceutical and human urine by capillary electrophoresis with electrochemiluminescence detection and studies on the pharmacokinetics. J Chromatogr B 879(13–14):871–877. doi:10.1016/j.jchromb.2011.02.032

45. Li YH, Wang CY, Sun JY, Zhou YC, You TY, Wang EK, Fung YS (2005) Determination of dioxopromethazine hydrochloride by capillary electrophoresis with electrochemiluminescence detection. Anal Chim Acta 550(1–2):40–46. doi:10.1016/j.aca.2005.06.045

46. Yin XB, Kang JZ, Fang LY, Yang XR (1055) Wang EK (2004) Short-capillary electrophoresis with electrochemiluminescence detection using porous etched joint for fast analysis of lidocaine and ofloxacin. J Chromatogr A 1–2:223–228. doi:10.1016/j.chroma.2004.09.001

47. Zhou X, Xing D, Zhu D, Tang Y, Jia L (2008) Development and application of a capillary electrophoresis-electrochemiluminescent method for the analysis of enrofloxacin and its metabolite ciprofloxacin in milk. Talanta 75(5):1300–1306. doi:10.1016/j.talanta.2008.01.040

48. Fang LY, Yin XB, Sun XH, Wang EK (2005) Determination of disopyramide in human urine by capillary electrophoresis with electrochemiluminescence detection of tris(2,2'-bipyridyl)ruthenium(II). Anal Chim Acta 537(1–2):25–30. doi:10.1016/j.aca.2005.01.020

49. Yuan J, Yin J, Wang E (2007) Characterization of procaine metabolism as probe for the butyrylcholinesterase enzyme investigation by simultaneous determination of procaine and its metabolite using capillary electrophoresis with electrochemiluminescence detection. J Chromatogr A 1154(1–2):368–372. doi:10.1016/j.chroma.2007.02.024

50. Peng X, Wang Z, Li J, Le G, Shi Y (2008) Electrochemiluminescence detection of clarithromycin in biological fluids after capillary electrophoresis separation. Anal Lett 41(7):1184–1199. doi:10.1080/00032710802052528

51. Zhao XC, You TY, Qiu HB, Yan JL, Yang XR, Wang EK (2004) Electrochemiluminescence detection with integrated indium tin oxide electrode on electrophoretic microchip for direct bioanalysis of lincomycin in the urine. J Chromatogr B 810(1):137–142. doi:10.1016/j.jchromb.2004.07.018

52. Liang Y-D, Song J-F, Xu M (2007) Electrochemiluminescence from successive electro- and chemo-oxidation of rifampicin and its application to the determination of rifampicin in pharmaceutical preparations and human urine. Spectrochim Acta, Part A 67(2):430–436. doi:10.1016/j.saa.2006.07.036

53. Chiu HY, Lin ZY, Tu HL, Whang C-W (2008) Analysis of glyphosate and aminomethylphosphonic acid by capillary electrophoresis with electrochemiluminescence detection. J Chromatogr A 1177(1):195–198. doi:10.1016/j.chroma.2007.11.042

54. Cai Q, Chen X, Qiu B, Lin Z (2011) Electrochemiluminescent detection method for glyphosate in soybean on carbon fiber-ionic liquid paste electrode. Chin J Chem 29(3):581–586

55. Wei W, Wei M, Cai Z, Liu S (2011) Determination of spectinomycin in human urine using ce coupled with electrogenerated chemiluminescence. Chromatographia 74(3–4):349–353. doi:10.1007/s10337-011-2060-0

56. Deng B, Lu H, Li L, Shi A, Kang Y, Xu Q (2010) Determination of the number of binding sites and binding constant between diltiazem hydrochloride and human serum albumin by ultrasonic microdialysis coupled with online capillary electrophoresis electrochemilumine-scence. J Chromatogr A 1217(28):4753–4756. doi:10.1016/j.chroma.2010.05.021

57. Sun XH, Liu JF, Cao WD, Yang XR, Wang EK, Fung YS (2002) Capillary electrophoresis with electrochemiluminescence detection of procyclidine in human urine pretreated by ion-exchange cartridge. Anal Chim Acta 470(2):137–145. doi:10.1016/s0003-2670(02)00780-8

58. Sun Y, Zhang Z, Xi Z, Shi Z (2009) Determination of naproxen in human urine by high-performance liquid chromatography with direct electrogenerated chemiluminescence detection. Talanta 79(3):676–680. doi:10.1016/j.talanta.2009.04.048

59. Cao WD, Yang XR, Wang EK (2004) Determination of reserpine in urine by capillary electrophoresis with electrochemiluminescence detection. Electroanalysis 16(3):169–174. doi:10.1002/elan.200402777

60. Sun H, Li L, Su M (2010) Simultaneous determination of proline and pipemidic acid in human urine by capillary electrophoresis with electrochemiluminescence detection. J Clin Lab Anal 24(5):327–333. doi:10.1002/jcla.20284

61. Yuan JP, Li T, Yin XB, Guo L, Jiang XZ, Jin WR, Yang XR, Wang EK (2006) Characterization of prolidase activity using capillary electrophoresis with tris(2,2′-bipyridyl)ruthenium(II) electrochemiluminescence detection and application to evaluate collagen degradation in diabetes mellitus. Anal Chem 78(9):2934–2938. doi:10.1021/ac051594x

62. Fu Z, Wang L, Wang Y (2009) Capillary electrophoresis-electrochemiluminescent detection of N, N-dimethyl ethanolamine and its application in impurity profiling and stability investigation of meclophenoxate. Anal Chim Acta 638(2):220–224. doi:10.1016/j.aca.2009.02.024

63. Yu C, Du H, You T (2011) Determination of imipramine and trimipramine by capillary electrophoresis with electrochemiluminescence detection. Talanta 83(5):1376–1380. doi:10.1016/j.talanta.2010.11.011

64. Sassolas A, Blum LJ, Leca-Bouvier BD (2009) Polymeric luminol on pre-treated screen-printed electrodes for the design of performant reagentless biosensors. Sens Actuators, B 139(1):214–221. doi:10.1016/j.snb.2009.01.020

65. Piao MH, Yang DS, Yoon KR, Lee SH, Choi SH (2009) Development of an Electrogenerated Chemiluminescence Biosensor using Carboxylic acid-functionalized MWCNT and Au Nanoparticles. Sensors 9(3):1662–1677

66. Xu ZA, Guo ZH, Dong SJ (2005) Electrogenerated chemiluminescence biosensor with alcohol dehydrogenase and tris(2,2′-bipyridyl)ruthenium (II) immobilized in sol-gel hybrid material. Biosens Bioelectron 21(3):455–461. doi:10.1016/j.bios.2004.10.032

67. Li M, Lee SH (2007) Analysis of monosaccharides by capillary electrophoresis with electrochemiluminescence detection. Anal Sci 23(11):1347–1349. doi:10.2116/analsci.23.1347

68. Dai H, Wu X, Xu H, Wang Y, Chi Y, Chen G (2009) A highly performing electrochemiluminescent biosensor for glucose based on a polyelectrolyte-chitosan modified electrode. Electrochim Acta 54(19):4582–4586. doi:10.1016/j.electacta.2009.03.042

69. Xiong Z-G, Li J-P, Tang L, Cheng Z-Q (2010) A novel electrochemiluminescence biosensor based on glucose oxidase immobilized on magnetic nanoparticles. Chin J Anal Chem 38(6):800–804. doi:10.3724/sp.j.1096.2010.00800

70. Zhu L, Li YX, Zhu GY (2002) A novel flow through optical fiber biosensor for glucose based on luminol electrochemiluminescence. Sens Actuators, B 86(2–3):209–214. doi:10.1016/s0925-4005(02)00173-9

71. Marquette CA, Blum LJ (1999) Luminol electrochemiluminescence-based fibre optic biosensors for flow injection analysis of glucose and lactate in natural samples. Anal Chim Acta 381(1):1–10. doi:10.1016/s0003-2670(98)00703-x

72. Martínez-Olmos A, Ballesta-Claver J, Palma A, Valencia-Mirón MDC, Capitán-Vallvey L (2009) A portable luminometer with a disposable electrochemiluminescent biosensor for lactate determination. Sensors 9(10):7694–7710

73. Lei R, Wang X, Zhu S, Li N (2011) A novel electrochemiluminescence glucose biosensor based on alcohol-free mesoporous molecular sieve silica modified electrode. Sens Actuators, B 158(1):124–129. doi:10.1016/j.snb.2011.05.054

74. Ballesta Claver J, Valencia Miron MC, Capitan-Vallvey LF (2009) Disposable electrochemiluminescent biosensor for lactate determination in saliva. Analyst 134(7):1423–1432. doi:10.1039/b821922b

75. Haghighi B, Bozorgzadeh S (2011) Fabrication of a highly sensitive electrochemiluminescence lactate biosensor using ZnO nanoparticles decorated multiwalled carbon nanotubes. Talanta 85(4):2189–2193. doi:10.1016/j.talanta.2011.07.071

76. Lin Z, Chen J, Chen G (2008) An ECL biosensor for glucose based on carbon-nanotube/Nafion film modified glass carbon electrode. Electrochim Acta 53(5):2396–2401. doi:10.1016/j.electacta.2007.09.063

77. Qiu B, Lin Z, Wang J, Chen Z, Chen J, Chen G (2009) An electrochemiluminescent biosensor for glucose based on the electrochemiluminescence of luminol on the nafion/glucose oxidase/poly(nickel(II)tetrasulfophthalocyanine)/multi-walled carbon nanotubes modified electrode. Talanta 78(1):76–80. doi:10.1016/j.talanta.2008.10.067

78. Chen Z, Wang J, Lin Z, Chen G (2007) A new electrochemiluminescent sensing system for glucose based on the electrochemiluminescent reaction of bis- 3,4,6-trichloro-2-(pentyloxycarbonyl)-phenyl oxalate. Talanta 72(4):1410–1415. doi:10.1016/j.talanta.2007.01.051

79. Li G, Lian J, Zheng X, Cao J (2010) Electrogenerated chemiluminescence biosensor for glucose based on poly(luminol-aniline) nanowires composite modified electrode. Biosens Bioelectron 26(2):643–648. doi:10.1016/j.bios.2010.07.003

80. Wang CY, Huang HJ (2003) Flow injection analysis of glucose based on its inhibition of electrochemiluminescence in a $Ru(bpy)_3^{2+}$-tripropylamine system. Anal Chim Acta 498(1–2):61–68. doi:10.1016/j.aca.2003.08.064

81. Marquette CA, Blum LJ (2003) Self-containing reactant biochips for the electrochemiluminescent determination of glucose, lactate and choline. Sens Actuators, B 90(1–3):112–117. doi:10.1016/s0925-4005(03)00046-7

82. Zhang L, Xu Z, Sun X, Dong S (2007) A novel alcohol dehydrogenase biosensor based on solid-state electrogenerated chemiluminescence by assembling dehydrogenase to Ru(bpy)(3)(2+)-Au nanoparticles aggregates. Biosens Bioelectron 22(6):1097–1100. doi:10.1016/j.bios.2006.03.026

83. Shan Y, Xu J–J, Chen H-Y (2010) Electrochemiluminescence quenching by CdTe quantum dots through energy scavenging for ultrasensitive detection of antigen. Chem Comm 46(28):5079–5081. doi:10.1039/c0cc00837k

84. Zheng XW, Guo ZH, Zhang ZJ (2001) Flow-injection electrogenerated chemiluminescence determination of epinephrine using luminol. Anal Chim Acta 441(1):81–86. doi:10.1016/s0003-2670(01)01090-x

85. Wang S, Yu J, Wan F, Ge S, Yan M, Zhang M (2011) Flow injection electrochemiluminescence determination of L-lysine using tris (2,2′-bipyridyl) ruthenium(II) $(Ru(bpy)_3^{2+}$ on indium tin oxide (ITO) glass. Anal Meth 3(5):1163–1167. doi:10.1039/c0ay00632g

86. Fang L, Lue Z, Wei H, Wang E (2008) Quantitative electrochemiluminescence detection of proteins: Avidin-based sensor and tris(2,2′-bipyridine) ruthenium(II) label. Biosens Bioelectron 23(11):1645–1651. doi:10.1016/j.bios.2008.01.023

87. Yin XB, Qi B, Sun XP, Yang XR, Wang EK (2005) 4-(Dimethylamino)butyric acid labeling for electrochemiluminescence detection of biological substances by increasing sensitivity with gold nanoparticle amplification. Anal Chem 77(11):3525–3530. doi:10.1021/ac0503198

88. Qiu B, Jiang X, Guo L, Lin Z, Cai Z, Chen G (2011) A highly sensitive method for detection of protein based on inhibition of $Ru(bpy)_3^{2+}$/TPrA electrochemiluminescent system. Electrochim Acta 56(20):6962–6965. doi:10.1016/j.electacta.2011.06.016

89. Brandon DL (2011) Detection of ricin contamination in ground beef by electrochemiluminescence immunosorbent assay. Toxins 3(4):398–408

90. Qolizadeh MR, Ebrahim K, Rahbar B, Karami E, Rostamkhany H, Musavi SH (2011) The effect of choline supplementation on the level of plasma free fatty acids and beta-hydroxybutyrate during a session of prolonged exercise. Annals Bio Res 2(6):253–260

91. Wei W, Kang X, Deng H, Lu Z, Jie Z (2011) Analysis of choline in milk powder using electrogenerated chemiluminescence including a mechanism study. Anal Lett 44(8):1381–1391. doi:10.1080/00032719.2010.512681

92. Deng S, Lei J, Cheng L, Zhang Y, Ju H (2011) Amplified electrochemiluminescence of quantum dots by electrochemically reduced graphene oxide for nanobiosensing of acetylcholine. Biosens Bioelectron 26(11):4552–4558. doi:10.1016/j.bios.2011.05.023

93. Tsafack VC, Marquette CA, Leca B, Blum LJ (1999) An electrochemiluminescence-based fibre optic biosensor for choline flow injection analysis. Analyst 125(1):151–155. doi:10.1039/a907709j

94. Sun L, Bao L, Hyun BR, Bartnik AC, Zhong YW, Reed JC, Pang DW, HcD Abrun, Malliaras GG, Wise FW (2008) Electrogenerated chemiluminescence from pbs quantum dots. Nano Lett 9(2):789–793

95. Dai H, Chi Y, Wu X, Wang Y, Wei M, Chen G (2009) Biocompatible electrochemiluminescent biosensor for choline based on enzyme/titanate nanotubes/chitosan composite modified electrode. Biosens Bioelectron 25(6):1414–1419. doi:101016/jbios200910042

96. Sassolas A, Blum LJ, Leca-Bouvier BD (2009) New electrochemiluminescent biosensors combining polyluminol and an enzymatic matrix. Anal Bioanal Chem 394(4):971–980. doi:10.1007/s00216-009-2780-2

97. Jin J, Muroga M, Takahashi F, Nakamura T (2010) Enzymatic flow injection method for rapid determination of choline in urine with electrochemiluminescence detection. Bioelectrochem 79(1):147–151. doi:10.1016/j.bioelechem.2009.12.005

98. Ballesta-Claver J, Diaz Ortega IF, Valencia-Miron MC, Capitan-Vallvey LF (2011) Disposable luminol copolymer-based biosensor for uric acid in urine. Anal Chim Acta 702(2):254–261. doi:10.1016/j.aca.2011.06.054

99. Chen Z, Zu Y (2008) Selective detection of uric acid in the presence of ascorbic acid based on electrochemiluminescence quenching. J Electroanal Chem 612(1):151–155. doi:10.1016/j.jelechem.2007.09.018

100. Waseem A, Yaqoob M, Nabi A, Greenway GM (2007) Determination of thyroxine using tris(2,2′-bipyridyl)ruthenium(III)-NADH enhanced electrochemiluminescence detection. Anal Lett 40(6):1071–1083. doi:10.1080/00032710701298495

101. Liu X, Ju H (2008) Coreactant enhanced anodic electrocherniluminescence of CdTe quantum dots at low potential for sensitive biosensing amplified by enzymatic cycle. Anal Chem 80(14):5377–5382. doi:10.1021/ac8003715

102. Li F, Pang YQ, Lin XQ, Cui H (2003) Determination of noradrenaline and dopamine in pharmaceutical injection samples by inhibition flow injection electrochemiluminescence of ruthenium complexes. Talanta 59(3):627–636. doi:10.1016/s0039-9140(02)00576-3

103. Zhao J, Chen M, Yu C, Tu Y (2011) Development and application of an electrochemiluminescent flow-injection cell based on CdTe quantum dots modified electrode for high sensitive determination of dopamine. Analyst 136(19):4070–4074. doi:10.1039/c1an15458c

104. Zhu LD, Li YX, Zhu GY (2002) Flow injection determination of dopamine based on inhibited electrochemiluminescence of luminol. Anal Lett 35(15):2527–2537. doi:10.1081/al-120016542

105. Xue L, Guo L, Qiu B, Lin Z, Chen G (2009) Mechanism for inhibition of/DBAE electrochemiluminescence system by dopamine. Electrochem Comm 11(8):1579–1582

106. Yu C, Yan J, Tu Y (2011) Electrochemiluminescent sensing of dopamine using CdTe quantum dots capped with thioglycolic acid and supported with carbon nanotubes. Microchim Acta 175(3–4):347–354. doi:10.1007/s00604-011-0666-4

107. Kang JZ, Yin XB, Yang XR, Wang EK (2005) Electrochemiluminescence quenching as an indirect method for detection of dopamine and epinephrine with capillary electrophoresis. Electrophoresis 26(9):1732–1736. doi:10.1002/elps.200410247

108. Yin X-B, Guo J-M, Wei W (2010) Dual-cloud point extraction and tertiary amine labeling for selective and sensitive capillary electrophoresis-electrochemiluminescent detection of auxins. J Chromatogr A 1217(8):1399–1406. doi:10.1016/j.chroma.2009.12.029

109. Guo W, Yuan J, Li B, Du Y, Ying E, Wang E (2008) Nanoscale-enhanced Ru(bpy)$_3^{2+}$ electrochemiluminescence labels and related aptamer-based biosensing system. Analyst 133(9):1209–1213. doi:10.1039/b806301j

110. Chang PL, Lee KH, Hu CC, Chang HT (2007) CE with sequential light-emitting diodeinduced fluorescence and electro-chemiluminescence detections for the determination of amino acids and alkaloids. Electrophoresis 28(7):1092–1099. doi:10.1002/elps.200600546

111. Zhuang YF, Ju HX (2005) Determination of reduced nicotinamide adenine dinucleotide based on immobilization of tris(2,2'-bipyridyl) ruthenium(II) in multiwall carbon nanotubes/Nafion composite membrane. Anal Lett 38(13):2077–2088. doi:10.1080/00032710500259441

112. Liu SC, Liu YJ, Li J, Guo ML, Pan W, Yao SZ (2006) Determination of mefenacet by capillary electrophoresis with electrochemiluminescence detection. Talanta 69(1):154–159. doi:10.1016/j.talanta.2005.09.020

113. Yin J, Guo W, Du Y, Wang E (2006) Facile separation and determination of Aconitine alkaloids in traditional Chinese medicines by CE with tris(2,2'-bipyridyl) ruthenium(II)-based electrochemi-luminescence detection. Electrophoresis 27(23):4836–4841. doi:10.1002/elps.200600288

114. Yin J, Xu Y, Li J, Wang E (2008) Analysis of quinolizidine alkaloids in *Sophora flavescens* Ait. by capillary electrophoresis with tris(2,2'-bipyridyl) ruthenium (II)-based electro-chemiluminescence detection. Talanta 75(1):38–42. doi:10.1016/j.talanta.2007.10.003

115. Chen Y, Lin Z, Chen J, Sun J, Zhang L, Chen G (2007) New capillary electrophoresis-electrochemiluminescence detection system equipped with an electrically heated Ru(bpy)$_3^{2+}$/multi-wall-carbon-nanotube paste electrode. J Chromatogr A 1172(1):84–91. doi:10.1016/j.chroma.2007.09.049

116. Deng B, Xie F, Li L, Shi A, Liu Y, Yin H (2010) Determination of galanthamine in Bulbus Lycoridis Radiatae by coupling capillary electrophoresis with end-column electrochemiluminescence detection. J Sep Sci 33(15):2356–2360. doi:10.1002/jssc.201000140

117. Gao Y, Xu Y, Han B, Li J, Xiang Q (2009) Sensitive determination of verticine and verticinone in Bulbus Fritillariae by ionic liquid assisted capillary electrophoresis-electrochemiluminescence system. Talanta 80(2):448–453. doi:10.1016/j.talanta.2009.07.012

118. Wang Z, Duan N, Hun X, Wu S (2010) Electrochemiluminescent aptamer biosensor for the determination of ochratoxin A at a gold-nanoparticles-modified gold electrode using N-(aminobutyl)-N-ethylisoluminol as a luminescent label. Anal Bioanal Chem 398(5):2125–2132. doi:10.1007/s00216-010-4146-1

119. Sun J, Du H, You T (2011) Determination of nicotine and its metabolite cotinine in urine and cigarette samples by capillary electrophoresis coupled with electrochemiluminescence. Electrophoresis 32(16):2148–2154. doi:10.1002/elps.201100075

120. Deng B, Ye L, Yin H, Liu Y, Hu S, Li B (2011) Determination of pseudolycorine in the bulb of lycoris radiata by capillary electrophoresis combined with online electrochemiluminescence using ultrasonic-assisted extraction. J Chromatogr B 879(13–14):927–932. doi:10.1016/j.jchromb.2011.03.002

121. Gao Y, Tian YL, Wang EK (2005) Simultaneous determination of two active ingredients in Flos daturae by capillary electrophoresis with electrochemiluminescence detection. Anal Chim Acta 545(2):137–141. doi:10.1016/j.aca.2005.04.071

122. Yuan B, Zheng C, Teng H, You T (2010) Simultaneous determination of atropine, anisodamine, and scopolamine in plant extract by nonaqueous capillary electrophoresis coupled with electrochemiluminescence and electrochemistry dual detection. J Chromatogr A 1217(1):171–174. doi:10.1016/j.chroma.2009.11.008

123. Li M, Lee SH (2007) Determination of trimethylamine in fish by capillary electrophoresis with electrogenerated tris(2,2′-bipyridyl)ruthenium(II) chemiluminescence detection. Luminescence 22(6):588–593. doi:10.1002/bio.1006

124. Guo Z, Gai P, Hao T, Duan J, Wang S (2011) Determination of malachite green residues in fish using a highly sensitive electrochemiluminescence method combined with molecularly imprinted solid phase extraction. J Agr Food Chem 59(10):5257–5262. doi:10.1021/jf2008502

125. Lo W-Y, Baeumner AJ (2007) Evaluation of internal standards in a competitive nucleic acid sequence-based amplification assay. Anal Chem 79(4):1386–1392

126. Lo W-Y, Baeumner AJ (2007) RNA internal standard synthesis by nucleic acid sequence-based amplification for competitive quantitative amplification reactions. Anal Chem 79(4):1548–1554

127. Li Y, Qi H, Fang F, Zhang C (2007) Ultrasensitive electrogenerated chemiluminescence detection of DNA hybridization using carbon-nanotubes loaded with tris(2,2′-bipyridyl) ruthenium derivative tags. Talanta 72(5):1704–1709

128. Lee J-G, Yun K, Lim G-S, Lee SE, Kim S, Park J-K (2007) DNA biosensor based on the electrochemiluminescence of $Ru(bpy)_3^{2+}$ with DNA-binding intercalators. Bioelectrochem 70(2):228–234

129. Chang Z, Zhou J, Zhao K, Zhu N, He P, Fang Y (2006) $Ru(bpy)_3^{2+}$-doped silica nanoparticle DNA probe for the electrogenerated chemiluminescence detection of DNA hybridization. Electrochim Acta 52(2):575–580

130. Miao W, Bard AJ (2004) Electrogenerated chemiluminescence. 77. DNA hybridization detection at high amplification with $[Ru(bpy)_3]^{2+}$-containing microspheres. Anal Chem 76(18):5379–5386. doi:10.1021/ac0495236

131. Wang H, Zhang C, Li Y, Qi H (2006) Electrogenerated chemiluminescence detection for deoxyribonucleic acid hybridization based on gold nanoparticles carrying multiple probes. Anal Chim Acta 575(2):205–211. doi:10.1016/j.aca.2006.05.080

132. Bertolino C, MacSweeney M, Tobin J, Neill B, Sheehan MM, Coluccia S, Berney H (2005) A monolithic silicon based integrated signal generation and detection system for monitoring DNA hybridisation. Biosens Bioelectron 21(4):565–573

133. Firrao G (2005) Detection of DNA/DNA hybridization by electrogenerated chemiluminescence. Int J Env Anal Chem 85(9–11):609–612

134. Hsueh YT, Collins SD, Smith RL (1998) DNA quantification with an electrochemiluminescence microcell. Sens Actuators, B 49(12):1–4

135. Zhang J, Qi H, Li Y, Yang J, Gao Q, Zhang C (2008) Electrogenerated chemiluminescence DNA biosensor based on hairpin DNA probe labeled with ruthenium complex. Anal Chem 80(8):2888–2894. doi:10.1021/ac701995g

136. Spehar A-M, Koster S, Kulmala S, Verpoorte E, de Rooij N, Koudelka-Hep M (2004) The quenching of electrochemiluminescence upon oligonucleotide hybridization. Luminescence 19(5):287–295

137. Wu A-H, Sun JJ, Zheng RJ, Yang HH, Chen GN (2010) A reagentless DNA biosensor based on cathodic electrochemiluminescence at a $C/C(x)O(1-x)$ electrode. Talanta 81(3): 934–940. doi:10.1016/j.talanta.2010.01.040

138. Wang X, He P, Fang Y (2010) A solid-state electrochemiluminescence biosensing switch for detection of DNA hybridization based on ferrocene-labeled molecular beacon. J Lumin 130(8):1481–1484. doi:10.1016/j.jlumin.2010.03.016

139. Spehar-Deleze A-M, Schmidt L, Neier R, Kulmala S, de Rooij N, Koudelka-Hep M (2006) Electrochemiluminescent hybridization chip with electric field aided mismatch discrimination. Biosens Bioelectron 22(5):722–729. doi:10.1016/j.bios.2006.02.013

140. Miao W, Bard AJ (2003) Electrogenerated chemiluminescence. 72. Determination of immobilized DNA and C-reactive protein on Au(111) electrodes using tris(2,2'-bipyridyl)ruthenium(II) labels. Anal Chem 75(21):5825–5834. doi:10.1021/ac034596v

141. Bruno JG, Parker JE, Holwitt E, Alls JL, Kiel JL (1998) Preliminary electrochemiluminescence studies of metal ion–bacterial diazoluminomelanin (DALM) interactions. J Biolumin Chemilumin 13(3):117–123

142. Boom R, Sol C, Weel J, Gerrits Y, de Boer M, Wertheim-van Dillen P (1999) A highly sensitive assay for detection and quantitation of human cytomegalovirus DNA in serum and plasma by PCR and electrochemiluminescence. J Clin Microbiol 37(5):1489–1497

143. Suzuki K, Yoshikawa T, Tomitaka A, Matsunaga K, Asano Y (2004) Detection of aerosolized varicella-zoster virus DNA in patients with localized herpes zoster. J Infec Dis 189. doi:10.1086/382029

144. Zhu D, Xing D, Shen X, Liu J, Chen Q (2004) High sensitive approach for point mutation detection based on electrochemiluminescence. Biosens Bioelectron 20(3):448–453

145. Tang YB, Xing D, Zhu DB, Liu JF (2006) An improved electrochemiluminescence polymerase chain reaction method for highly sensitive detection of plant viruses. Anal Chim Acta 582:275

146. Liu J, Xing D, Shen X, Zhu D (2005) Electrochemiluminescence polymerase chain reaction detection of genetically modified organisms. Anal Chim Acta 537(12):119–123

147. Zhang L, Schwartz G, O'Donnell M, Harrison RK (2001) Development of a novel helicase assay using electrochemiluminescence. Anal Biochem 293(1):31–37

148. Yang ML, Liu CZ, Qian KJ, He PG, Fang YZ (2002) Study on the electrochemiluminescence behavior of ABEI and its application in DNA hybridization analysis. Analyst 127(9):1267–1271. doi:10.1039/b205783b

149. Calvo-Munoz ML, Dupont-Filliard A, Billon M, Guillerez S, Bidan G, Marquette C, Blum L (2005) Detection of DNA hybridization by ABEI electrochemiluminescence in DNA-chip compatible assembly. Bioelectrochem 66(1–2):139–143. doi:10.1016/j.bioelechem.2004.04.009

150. Spehar-Deleze AM, Suomi J, Jiang Q, De Rooij N, Koudelka-Hep M, Kulmala S (2006) Heterogeneous oligonucleotide-hybridization assay based on hot electron-induced electrochemiluminescence of a rhodamine label at oxide-coated aluminum and silicon electrodes. Electrochim Acta 51(25):5438–5444. doi:10.1016/j.electacta.2006.02.017

151. Schmittel M, Lin H-W (2007) Quadruple-channel sensing: a molecular sensor with a single type of receptor site for selective and quantitative multi-ion analysis. Angew Chem Intl Ed 46(6):893–896

152. Dennany L, Forster RJ, White B, Smyth M, Rusling JF (2004) Direct electrochemiluminescence detection of oxidized DNA in ultrathin films containing [Os(bpy)$_2$(PVP)$_{10}$]$^{2+}$. J Am Chem Soc 126(28):8835–8841

153. Li Y, Qi H, Peng Y, Yang J, Zhang C (2007) Electrogenerated chemiluminescence aptamer-based biosensor for the determination of cocaine. Electrochem Comm 9(10):2571–2575

154. Ye S, Li H, Cao W (2011) Electrogenerated chemiluminescence detection of adenosine based on triplex DNA biosensor. Biosens Bioelectron 26(5):2215–2220. doi:10.1016/j.bios.2010.09.037

155. Wang X, Dong P, He P, Fang Y (2010) A solid-state electrochemiluminescence sensing platform for detection of adenosine based on ferrocene-labeled structure-switching signaling aptamer. Anal Chim Acta 658(2):128–132. doi:10.1016/j.aca.2009.11.007

156. Chen L, Cai Q, Luo F, Chen X, Zhu X, Qiu B, Lin Z, Chen G (2010) A sensitive aptasensor for adenosine based on the quenching of Ru(bpy)$_3$$^{2+}$-doped silica nanoparticle ECL by ferrocene. Chem Comm 46(41):7751–7753. doi:10.1039/c0cc03225e

157. Yao W, Wang L, Wang H, Zhang X, Li L (2009) An aptamer-based electrochemiluminescent biosensor for ATP detection. Biosens Bioelectron 24(11):3269–3274. doi:10.1016/j.bios.2009.04.016

158. Liao Y, Yuan R, Chai Y, Mao L, Zhuo Y, Yuan Y, Bai L, Yuan S (2011) Electrochemiluminescence quenching via capture of ferrocene-labeled ligand-bound

aptamer molecular beacon for ultrasensitive detection of thrombin. Sens Actuators, B 158(1):393–399. doi:10.1016/j.snb.2011.06.045

159. Shan Y, Xu JJ, Chen HY (2011) Enhanced electrochemiluminescence quenching of CdS:Mn nanocrystals by CdTe QDs-doped silica nanoparticles for ultrasensitive detection of thrombin. Nanoscale 3(7):2916–2923. doi:10.1039/c1nr10175g

160. Zhang J, Chen P, Wu X, Chen J, Xu L, Chen G, Fu F (2011) A signal-on electrochemiluminescence aptamer biosensor for the detection of ultratrace thrombin based on junction-probe. Biosens Bioelectron 26(5):2645–2650. doi:10.1016/j.bios.2010.11.028

161. Xu Y, Dong P, Zhang X, He P, Fang Y (2011) Solid-state electrochemiluminescence protein biosensor with aptamer substitution strategy. Sci China-Chem 54(7):1109–1115. doi:10.1007/s11426-011-4278-y

162. Wang X, Dong P, Yun W, Xu Y, He P, Fang Y (2010) Detection of T4 DNA ligase using a solid-state electrochemiluminescence biosensing switch based on ferrocene-labeled molecular beacon. Talanta 80(5):1643–1647. doi:10.1016/j.talanta.2009.09.060

163. Liang S, Connel GJ (2009) An electrochemiluminescent aptamer switch for a high-throughput assay of an RNA editing reaction. RNA 15(10):1929–1938. doi:10.1261/rna.1720209

164. Gott JM (2000) Emeson RB Functions and mechanisms of RNA editing. Annu Rev Genet 34:499–531

165. Deaver DR (1995) A new non-isotopic detection system for immunoassays. Nature 377(6551):758–760

166. Henchal EA, Teska JD, Ludwig GV, Shoemaker DR, Ezzell JW (2001) Current laboratory methods for biological threat agent identification. Clin Lab Med 21(3):661–678

167. Higgins JA, Ibrahim MS, Knauert FK, Ludwig GV, Kijek TM, Ezzell JW, Courtney BC, Henchal EA (1999) Sensitive and rapid identification of biological threat agents. Ann N Y Acad Sci 894:130–148

168. Bruno JG, Kiel JL (2002) Use of magnetic beads in selection and detection of biotoxin aptamers by electrochemiluminescence and enzymatic methods. Biotechniques 32(1):178–180

169. Filella X, Friese S, Roth HJ, Nussbaum S, Wehnl B (2000) Technical performance of the Elecsys CA 72-4 test–development and field study. Anticancer Res 20(6D):5229–5232

170. Stieber P, Molina R, Chan DW, Fritsche HA, Beyrau R, Bonfrer JM, Filella X, Gornet TG, Hoff T, Jager W, van Kamp GJ, Nagel D, Peisker K, Sokoll LJ, Troalen F, Untch M, Domke I (2001) Evaluation of the analytical and clinical performance of the Elecsys CA 15–3 immunoassay. Clin Chem 47(12):2162–2164

171. Hubl W, Chan DW, Van Ingen HE, Miyachi H, Molina R, Filella X, Pitzel L, Ruibal A, Rymer JC, Bagnard G, Domke I (1999) Multicenter evaluation of the elecsys CA 125 II assay. Anticancer res 19(4A):2727–2733

172. van Ingen HE, Chan DW, Hubl W, Miyachi H, Molina R, Pitzel L, Ruibal A, Rymer JC, Domke I (1998) Analytical and clinical evaluation of an electrochemiluminescence immunoassay for the determination of CA 125. Clin Chem 44(12):2530–2536

173. Baeumner AJ, Humiston MC, Montagna RA, Durst RA (2001) Detection of viable oocysts of cryptosporidium parvum following nucleic acid sequence based amplification. Anal Chem 73(6):1176–1180

174. Kuczynska E, Boyer DG, Shelton DR (2003) Comparison of immunofluorescence assay and immunomagnetic electrochemiluminescence in detection of *Cryptosporidium parvum* oocysts in karst water samples. J Microbiol Methods 53(1):17–26

175. Weinreb PH, Yang WJ, Violette SM, Couture M, Kimball K, Pepinsky RB, Lobb RR, Josiah S (2002) A cell-free electrochemiluminescence assay for measuring beta1-integrin-ligand interactions. Anal Biochem 15 306(2):305–313

176. Debad JD, Glezer EN, Wohlstadter J, Sigal GB, Leland JK (2004) In: Bard AJ (ed) Electrogenerated chemiluminescence. Marcel Dekker, New York, p 359

177. Bruno JG, Yu H (1996) Immunomagnetic-electrochemiluminescent detection of bacillus anthracis spores in soil matrices. App Envir Microbiol 62(9):3474–3476

178. Yu H (1997) Enhancing immunoassay possibilities using magnetic carriers in biological fluids. In: Optical diagnostics of biological fluids and advanced techniques in analytical cytology. SPIE, The international society for optical engineering, Bellingham, WA, pp 168–179

179. Yu H, Bruno JG (1996) Immunomagnetic-electrochemiluminescent detection of *Escherichia coli* O157 and *Salmonella typhimurium* in foods and environmental water samples. Appl Env Microbiol 62:587

180. Crawford CG, Wijey C, Fratamico P, Tu SI, Brewster J (2000) Immunomagnetic-electrochemiluminescent detection of *E. Coli* O157:H7 in ground beef. J Rapid Methods Autom Microbiol 8(4):249–264

181. Miao W, Bard AJ (2003) Electrogenerated chemiluminescence. 72. Determination of immobilized DNA and C-reactive protein on Au(111) electrodes using tris(2,2′-bipyridyl) ruthenium(II) labels. Anal Chem 75(21):5825–5834

182. Lee YM, Johnson PW, Call JL, Arrowood MJ, Furness BW, Pichette SC, Grady KK, Reeh P, Mitchell L, Bergmire-Sweat D, Mackenzie WR (2001) Tsang VC (2001) Development and application of a quantitative, specific assay for *Cryptosporidium parvum* oocyst detection in high-turbidity environmental water samples. Am J Trop Med Hyg 65(1):1–9

183. Call JL, Arrowood M, Xie LT, Hancock K, Tsang VC (2001) Immunoassay for viable Cryptosporidium parvum oocysts in turbid environmental water samples. J Parasitol 87(1):203–210

184. Yu H (1996) Enhancing immunoelectrochemiluminescence (IECL) for sensitive bacterial detection. J Immunol Methods 10;192(1–2):63–71

185. Kijek TM, Rossi CA, Moss D, Parker RW, Henchal EA (2000) Rapid and sensitive immunomagnetic-electrochemiluminescent detection of staphyloccocal enterotoxin B. J Immunol Methods 236(1–2):9–17

186. Yu H, Raymonda JW, McMahon TM, Campagnari AA (2000) Detection of biological threat agents by immunomagnetic microsphere-based solid phase fluorogenic- and electro-chemiluminescence. Biosens Bioelectron 14(10):829–840

187. Gatto-Menking DL, Yu H, Bruno JG, Goode MT, Miller M, Zulich AW (1995) Sensitive detection of biotoxoids and bacterial spores using an immunomagnetic electrocheminescence sensor. Biosen Bioelectron 10(67):501–507

188. Bruno JG, Kiel JL (1999) In vitro selection of DNA aptamers to anthrax spores with electrochemiluminescence detection. Biosens Bioelectron 14(5):457–464

189. Pyati R, Richter MM (2007) ECL-Electrochemical luminescence. Ann Rep Sec "C" (Phy Chem) 103(0):12–78

190. Zhao J, Chen M, Yu C, Tu Y (2011) Development and application of an electrochemiluminescent flow-injection cell based on CdTe quantum dots modified electrode for high sensitive determination of dopamine. Analyst 136(19):4070–4074. doi:101039/c1an15458c